What people say about this book

"This is a book of incontestable importance. It shows how an industry widely assumed to be benign is responsible for misery, destitution and destruction. Whether we like it or not, all of us are participants in its abuses. *Pulping the South* is a wake-up call to the world."

George Monbiot, author of *No Man's Land*

"This incisive study lays bare the anatomy of the plantations business. Packed with detail and backed by crystal-clear analysis, this book is an essential tool for all who support the social movements of the South that are resisting this theft of land and livelihood."

Marcus Colchester, Director, Forest Peoples Programme

"This book provides exhaustive and provocative scrutiny of the social and environmental impacts of the expansion of the pulp and paper industry in the South. In view of the increasing local social conflicts surrounding the appropriation and use of land and water resources by the industry, the current international discussion on biodiversity protection and indigenous people's rights, and the growing demand for pulp and paper products in developing countries, the book is a most timely addition to the literature. It will make valuable reading for scholars, environmentalists, peoples' associations, national and international policy makers, as well as those involved in the industry itself."

Krishna Ghimire, Project Leader, United Nations Research Institute for Social Development, Geneva

About the authors

Ricardo Carrere is a former professional forester with a background in eucalyptus plantations. He is currently deputy director of the Third World Institute in Uruguay, which is a non-governmental organization working on environment and development issues internationally. He is also the regional coordinator of the Latin American and Caribbean NGO Network on Multilateral Development Banks and a member of the Steering Committee of the World Rainforest Movement. Since 1988 he has concentrated his energies on research and campaigning at national and international levels on both native forests and commercial tree plantations. He is the author of numerous books in Spanish as well as, in English, *Forestry Geopolitics and Sustainable Development: An Overview and the Case of Uruguay* (Montevideo, CIEDUR, 1992).

Larry Lohmann spent much of the 1980s working with voluntary organizations in Thailand on social and environmental issues. He has published in scholarly and environmental journals and is co-editor (with Marcus Colchester) of *The Struggle for Land and the Fate of the Forests* (Zed, 1993) and co-author (with Simon Fairlie, Nicholas Hildyard and Sarah Sexton) of *Whose Common Future? Reclaiming the Commons* (Earthscan, 1993). From 1990 to 1994 he was an Associate Editor of *The Ecologist*.

Pulping the South

Industrial Tree Plantations and the World Paper Economy

Ricardo Carrere and Larry Lohmann

Zed Books Ltd

London and New Jersey

Pulping the South was first published by Zed Books Ltd,
7 Cynthia Street, London N1 9JF, UK and
165 First Avenue, Atlantic Highlands, New Jersey 07716, USA,
in association with World Rainforest Movement, 228 Macalister
Road, 10400 Penang, Malaysia and 8 Chapel Row, Chadlington,
Oxfordshire, OX7 3NA, UK, in 1996.

Published in Burma, Cambodia, Laos and Thailand
by White Lotus Co. Ltd., GPO Box 1141, Bangkok 10501, Thailand.

Cover designed by Andrew Corbett.
Cover photograph: Eucalyptus plantation, Brazil
© David Reed/Panos Pictures.
Typeset in Baskerville by Larry Lohmann.
Printed and bound in the United Kingdom
by Redwood Books Ltd, Trowbridge.

A catalogue record for this book
is available from the British Library.

Library of Congress Cataloging-in-Publication Data

Carrere, Ricardo.
  Pulping the South: Industrial tree plantations and the world paper
economy / Ricardo Carrere and Larry Lohmann.
    p. cm.
  Includes bibliographical references and index.
  ISBN 1-85649-437-3 (cl). — ISBN 1-85649-438-1 (p)
  1. Pulpwood crops—Southern Hemisphere. 2. Tree Farms—Southern
Hemisphere. 3. Paper Industry. I. Lohmann, Lawrence. II. Title.
SD356.35.S68C37   1996
338.1'74983'091814—dc20   96-19820
                                                                CIP

ISBN 1-85649-437-3 hb
ISBN 1-85649-438-1 pb
White Lotus ISBN 974 8496805

Contents

Acknowledgements   viii

PART ONE: PLANTATIONS, PULP AND PAPER

1. Introduction   3
     Commercial plantations and forests   3
     The advent of large-scale monocultures   5
     Plantation extent   6
     Plantation imperialism   8
     About this book   11
     Looking beyond this book   12

2. Introducing Pulp and Paper   16
     The evolution of a wood-based industry   16
     From wood to pulp   19
     From pulp to paper   21
     From paper back to pulp   23
     Large scale, capital intensity and centralization   27
     Boom and bust   30
     Concentration and liberalization   33
     Economic risks in South and North   34
     Consumption and demand creation   36

3. Emergence of a Global System   41
     A global market   41
     Export destinations   46
     Causes of globalization   47
     Shifting pulp production   51
     Dynamics of globalization: the case of Japan   54

4. Impacts on People and their Environment   60
     Unfair to tree plantations?   60
     The two libraries   61
     Impacts on water   63
     Impacts on biodiversity   69
     Impacts on the soil   75
     Industrial pollution   80
     Other socioeconomic impacts   82
     Conclusions   86

5. Actors behind the Scenes 87
   Pulp and paper firms 87
   Consultancy companies 89
   Technology suppliers 90
   Industry associations and alliances 91
   Bilateral agencies 93
   State investment and export credit agencies 95
   Multilateral agencies 97
   National governments 101
   Research institutes and NGOs 103
   Conclusion 104

6. Managing Resistance 114
   Non-threatening resistance 114
   More difficult forms of resistance 118
   Public relations, intelligence and 'astroturf' 122
   Supporting conditions 126
   Dividing experts from plantation opponents 127
   Stories for the uninformed public 131
   Conclusion 144

PART TWO: TREE PLANTATIONS IN THE SOUTH

7. Brazil: The Eucalyptus Pulp Giant 147
   An apparent success 148
   Aracruz 149
   Bahia Sul Celulose 155
   CENIBRA 159
   Jari and Companhia Florestal Monte Dourado 162
   Riocell 165
   Other planned projects 167
   Conclusions 168

8. Chile: A Plantation Model Imposed by a Dictatorship 169
   The native forests 169
   Genesis of the current model 171
   Socioeconomic and environmental consequences 175
   Environmental degradation 180
   Conclusions 185

9. Uruguay: 'Forests' on the Grasslands 186
   From tree planting to industrial plantations 186
   Domestic conditions encouraging the industrial model 188
   External conditions 189
   Voluntary blindness 192
   A bad investment for the nation 195
   Conclusions 197

10. South Africa: A Fibre Exporter with Few Forests   **198**
    A lack of firewood in a sea of trees   201
    More power to the powerful   201
    Jobs, work and migration   202
    Changes in rural lifestyles   204
    Water: a scarce and disputed resource   205
    Biodiversity in danger   207
    Soil degradation   209
    Conclusions   209

11. Indonesia: Deforestation and Repression   **211**
    Concentration   211
    Export focus   213
    Foreign involvement   214
    Plantations and deforestation   219
    Further subsidies   222
    Disintegration and resistance   224

12. Thailand: From 'Reforestation' to Contract Farming   **229**
    Manufacturing and export growth   229
    Official collaboration   231
    Foreign subsidies   233
    Popular resistance and its effects   235
    The struggles of the 1990s   237
    From planting leased land to contract farming   239
    Moving abroad   240
    The Thai Forestry Sector Master Plan   242

13. Conclusion: Looking to the Future   **246**
    The priority of politics   246
    Working positions   250
    Alliances among interest groups   250
    International solidarity   251

Bibliography   **254**

Index   **273**

Acknowledgements

This book, commissioned by the World Rainforest Movement at its meeting in Delhi in April 1994, has its origins in increasing concern among non-governmental organizations in the South over the spread of monoculture tree plantations. It is intended as a tool for all movements alarmed at the social, political and environmental effects of these plantations.

The authors, both of whom have been long involved in the plantations issue, divided their work on this book equally and were in constant consultation with each other across the Atlantic throughout its writing. Ricardo Carrere, based at the Instituto del Tercer Mundo in Uruguay, was responsible for drafting Chapters 4, 7, 8, 9 and 10; Larry Lohmann, based in the UK, for Chapters 2, 3, 5, 11 and 12. Chapters 1, 6 and 13 were jointly written.

The authors have been helped with information, advice, translations and criticism by more people than it is possible to name. Among those to whom thanks are due are Chris Albertyn, Patrick Anderson, Ndinga Assitou, Bill Barclay, Isabel Bermejo, Teresa Brooks, Raymond Bryant, Owen Cameron, Chee Yoke Ling, Marcus Colchester, Saliem Fakir, Anna Fanzeres, Maurizio Farhan, David Fig, Mark Gandar, Kate Geary, Alastair Graham, Helen Groome, Ted Gutman, David Hallowes, John Hanson, Nicholas Hildyard, S. R. Hiramath, Tomoya Inyaku, Sonoko Kawakami, Edda Kirleis, José Koopmans, Yoichi Kuroda, Sari Kuvaja, Chris Lang, Marvic Leonen and the staff of the Legal Rights and Natural Resources Center, Karin Lindahl, M. Patricia Marchak, Francesco Martone, Sarah Mason, Aubrey Mayer, Francisco Menezes, Moema Miranda, Robert Molteno, Sandra Moniaga, Roger Olsson, Juan-Pablo Orrego, David Orton, Saskia Ozinga, Ian Penna, Prompana Kuaicharoen, Noel Rajesh, Ulf Rasmusson, Sarah Roberts, Anne Rodford, Grant Rosoman, Sarah Sexton, Vandana Shiva, Pam Simmons, Maureen Smith, David Sonnenfeld, Srisuwan Kuankachorn, Antonio Thomen, Rowan Tilly, Marko Ulvila, Ann Danaiya Usher, Hernán Verscheure, Thomas Wallgren, Jeremy Whitham, Alex Wilks, Wirawat Theeraprasat, Witoon Permpongsacharoen, Al Wong, and Roger Wright. None of these people, of course, would necessarily agree with all or any of the conclusions reached in the book.

The authors appreciate the financial support of the Heinrich Böll Foundation, NOVIB, and IDRC, without which this book could not have been written.

Part One
Plantations, Pulp and Paper

Chapter 1

Introduction

To millions of people across the world today, the pulp and paper industry is a growing problem. The chipping of native forests to provide raw material for the industry is being opposed bitterly by local people and environmentalists from Australia to Finland, and from Chile to Canada (WALHI and YLBHI 1992, Hamilton 1995, PRS 1994, MacIsaac and Champagne 1994, WCWC 1994, Olsson 1995). Also widespread are protests at pulp mill pollution which has sucked oxygen from rivers, ruined fisheries and drinking water supplies, and increased the burden of highly-toxic chlorinated organic compounds in animal and human bodies (*BP* 12.7.1995, *EBY* 13.3, Greenpeace International 1994).

This book is concerned with a third activity of the pulp and paper industry — one which is often less well-publicized and which, at first glance, might seem more benign: planting trees. To help feed pulp and paper mills, vast monocultures of conifers, eucalyptus, acacia, and other species are being established both in the North and, increasingly, in the South, where fast tree growth, inexpensive land and labour, and lavish subsidies combine to make wood especially cheap. As swatches of exotic trees invade native woodlands, grasslands, farmlands and pastures, the results, in country after country, have been impoverishment, environmental degradation, and rural strife.

In documenting the often-hidden record of industrial pulpwood plantations in the South and what lies behind them, this book hopes to contribute to new ways of thinking about one of the world's most important industries as it undergoes rapid globalization.

Commercial plantations and forests

Plantations, like forests, are full of trees. But the two are usually radically different. A forest is a complex, self-regenerating system, encompassing soil, water, microclimate, energy, and a wide variety of plants and animals in mutual relation. A commercial plantation, on the other hand, is a cultivated area whose species and structure have been simplified dramatically to produce only a few goods, whether lumber, fuel, resin, oil, or fruit. A plantation's trees, unlike those of a forest, tend to be of a small range of species and ages, and to require extensive and continuing human intervention.

The distinction, of course, is not always hard and fast. A 'native forest' where economically unimportant species have been eliminated may wind up as simplified, and as in need of constant human maintenance to stay that way, as any plantation. Much of Europe's 'forest' falls into this category. On the other hand, some diverse, seemingly 'natural' forests either started out as plantations, having then been abandoned, or continue to be carefully 'cultivated' by local people, as in Kayapo-inhabited areas in Brazil (Posey 1985, 1990).

The industrial monocrops with which this book is concerned have a much less ambiguous status. Resulting from aggressive, thoroughgoing transformation of a landscape, they are much closer to an industrial agricultural crop than to either a forest as usually understood, a traditionally-managed coppice, or a traditional farm plot. Usually consisting of thousands or even millions of trees of the same species, bred for rapid growth, uniformity and high yield of raw material, and planted in even-aged stands, they require intensive preparation of the soil, fertilization, regular spacing of trees, seedling selection, intensive weeding using machines or herbicides, use of pesticides, thinning, mechanized harvesting, and, in some cases, pruning. Such plantations may be established either on single parcels of land owned or rented by a company or on a collection of smallholdings.

Even many 'non-industrial' plantations are today being established on this industrial model. In some places, for instance, large-scale rapid-growth monocrops are being grown on the false assumption that they can 'protect' water catchment areas or soils in the way a native forest would. Other extensive monocrops, often of exotic species, are being established with the stated purpose of providing fuelwood to local people. Industrial-type plantations are also being promoted as a way of absorbing emissions of carbon dioxide which lead to global warming: companies or countries are held to have 'compensated' for their heavy CO_2 emissions in one place if they plant swatches of fast-growing trees in another. Such plantations, even if not aimed at wood production, can have important impacts on industrial wood prices.

In contrast to such plantations — which are organized to meet one or two needs of large-scale manufacturing concerns or other centralizing actors — are attempts to plant trees in ways responsive to a wide variety of interlocked local concerns. In some agroforestry systems, for example, a diversity of trees are chosen and planted to provide protection, shade and food for livestock, fruit and wood for humans, and protection, nutrients and water for crops, thus helping to keep production diverse and in harmony with local landscapes and needs (Groome 1991, Shiva and Bandyopadhyay 1987, Shiva 1991b).

Another useful contrast to the industrial plantation model with which this book is concerned is offered by efforts to restore degraded forests or woodlands by planting trees of some or all of the original species. Here the objective is not to produce large quantities of whatever wood is suitable for industrial markets, but to restore diverse ecosystems using native species. Thus a eucalyptus tree, when planted in one of its native regions in Australia as a way of helping to regenerate an earlier ecosystem in ways approved by local people, may be considered to be a contribution to reforestation. The same tree, when planted as part of a large-scale pulpwood monocrop in India or Uruguay, is not only not a contribution to 'reforestation', but is likely to contribute to environmental degradation and social problems. Planting a tree, whether native or exotic, is in itself neither a positive or a negative process. It is the social and geographical structures within which that tree is planted which make it one or the other.

The advent of large-scale monocultures

Historically, tree plantations have most commonly consisted of fruit-bearing species such as olives, figs, date palms, and apples, as well as a diversity of trees cultivated for fodder, shelter, medicine and resin, or for aesthetic or religious reasons. Although teak and eucalyptus began to be grown in Asia, Africa and Latin America in the 19th century, extensive plantations of trees suitable for industry are mainly a 20th-century phenomenon, having generally been established as a result of overexploitation of native forests for wood. Such plantations are now expanding faster than ever, nowhere more quickly than in the South. Between 1965 and 1980, tree plantation area in the tropics tripled (Evans 1991), and between 1980 and 1990 increased again between two and three times (Pandey 1992, World Bank 1994b, Evans 1992).

Although plantations are promoted for a variety of reasons — to hold back desertification, to feed sawmills, to provide fuelwood, to diversify agricultural production — the trees most often planted today are fast-growing species favoured by industry for paper pulp or other low-grade wood products. In 1980, tropical plantations were estimated to consist of over 70 per cent eucalyptus and pines (see Table 1.1) (Evans 1991), and the proportion is almost certainly even higher today. Pine and eucalyptus are also widespread in the non-tropical regions of Argentina, Chile, Uruguay, China, South Africa, Australia, New Zealand, the southern United States, Spain, and Portugal. It is often the case that only one species will dominate a particular country's fast-growing tree plantations — for example, *Pinus radiata* in Chile and New Zealand, *Eucalyptus grandis* in Uruguay and Brazil, and *E. camaldulensis* in Thailand.

TABLE 1.1
Species used in tropical tree plantations

Genus/group	Species	Per cent
Eucalyptus	*E. grandis, camaldulensis, globulus, saligna, tereticornis, robusta, citriodora, urophylla, deglupta,* other	38
Pine	*Pinus patula, caribaea, elliotti, merkusii, kesiya, oocarpa,* other	34
Teak	*Tectona grandis*	14
Other hardwoods	*Acacia, Gmelina, Leucaena, Grevillea, Meliacaea, Terminalia, Albizzia, Prosopis, Casuarina, Cordia, Triplochiton,* other	12
Non-pine conifers	*Araucaria cunninghami, A. angustifolia, Cupressus lusitanica,* other	3

Source: Evans 1992

Plantation extent

Due to different reporting methods and shortages of data, no reliable estimates exist of the global extent of tree plantations. Table 1.2 attempts to suggest the rough extent of tree plantations in *tropical* countries in 1990 by comparing the figures contained in two serious studies, those of Evans (1992) and Pandey (1992). Included are country estimates by Pandey of the extent of plantations of species often used for pulpwood. Evans calculates that tropical tree plantations covered 42.7 million hectares in 1990 and Pandey 43.9 million hectares by the end of that year. The World Bank's (1994b) estimate for 1990, meanwhile, is 37.5 million hectares.

Michael D. Bazett's estimate that specifically *industrial* plantations occupy 99.3 million hectares in both tropical and non-tropical regions is contained in Table 1.3. Table 1.4, finally, presents Bazett's rough figures for *fast-growing* industrial plantations — those yielding 12 cubic metres or more per year per hectare — in the late 1980s. According to two sources, over 19 million hectares of fast-growing pine plantations and over six million hectares of eucalyptus are currently in existence (Bazett 1993, Wilson 1991); another claims that eucalyptus plantations total more than ten million hectares and acacia plantations 3.4 million

TABLE 1.2
Area covered by tree plantations in the tropics, 1990, thousand ha, selected countries

COUNTRY	EVANS	PANDEY			
	Total	Total	euca.	pine	acacia
Africa					
Angola	195	172	135	20	-
Burkina Faso	46	28	7	-	2
Burundi	80	132	40	8	25
Congo	18	53	35	15	-
Ethiopia	270	270	95	15	-
Ghana	76	75	14	1	-
Kenya	190	168	17	75	-
Madagascar	295	310	130	150	-
Malawi	156	180	30	75	5
Mozambique	45	40	14	23	-
Nigeria	259	216	11	-	-
Rwanda	110	125	60	44	-
Senegal	145	160	40	-	18
Sudan	330	290	23	-	145
Tanzania	100	220	25	45	10
Zaire	56	60	20	2	4
Zambia	65	68	26	40	-
Zimbabwe	125	120	20	80	16
Asia and Oceania					
Bangladesh	380	335	-	-	-
Fiji	85	104	-	40	-
India	14,000	18,900	4,800	40	3,000
Indonesia	3,700	8,750	-	600	75
Malaysia	90	116	8	7	75
New Caledonia	23	10	-	10	-
PNG	44	43	10	8	-
Pakistan	102	240	-	-	-
Philippines	100	290	-	-	-
Sri Lanka	195	198	45	31	-
Thailand	560	776	62	92	-
Viet Nam	616	2,100	245	400	-
The Americas					
Bolivia	37	40	-	-	-
Brazil	7,150	7,000	3,617	2,090	-
Colombia	250	180	31	88	-
Costa Rica	40	40	10	15	-
Cuba	316	350	35	160	-
Ecuador	60	64	44	13	-
Mexico	263	155	38	62	-
Nicaragua	38	20	-	13	-
Peru	272	263	211	13	-
Venezuela	350	362	70	245	-

hectares (Hagler 1995).While, according to Bazett, fast-growing plantations amount to only about a quarter of the total industrial plantation area, their importance to global wood supply, particularly pulpwood supply, is out of proportion to their size.

TABLE 1.3
Area under industrial plantations (million ha)

Region	Coniferous	Non-coniferous
Asia	31.6	8.2
Europe	15.7	3.3
Former USSR	16.9	-
North America	12.0	0.5
Latin America	3.5	2.9
Africa	1.3	1.2
Oceania	2.1	0.1
TOTAL	83.1	16.2

Source: Bazett 1993

Planting is expected to continue at a high rate. According to the World Bank (1994), an area of 10 to 12 million hectares is planted annually, half in China, although how much of this survives is unclear. India's eighth Five-Year Plan called for 17 million hectares of plantations, and Brazil has set out a programme for 12 million hectares of plantation by the year 2000. In Indonesia, some industrialists expect three to 4.6 million hectares of land to be under short-rotation pulpwood plantations by the year 2003 (Bazett 1993, Soetikno 1993), while Thai officials envisage over four million hectares being put under private-sector plantations by 2020. Ethiopia, meanwhile, has mooted plans to plant as much as 3.5 million hectares by 2000, Malaysia 500,000, and Burundi 300,000 (Evans 1991).

Plantation imperialism

The small range of genuses and species used in industrial plantations, as well as the nearly identical forestry development plans by which they are promoted, reflect a long-established and deeply-rooted forestry imperialism. By and large, plans for large-scale plantations do not arise in answer to the diversity of local needs. Nor, as this book attempts to show, do they often meet those needs. Rather, they function in a way which responds disproportionately to the needs of a Northern-dominated industrial economy while also serving bureaucratic ends.

TABLE 1.4
Area of fast-growing plantations at the end of the 1980s
(thousand ha)

Location	Coniferous	Non-coniferous	Total
Southern US	12,000★	500	12,500
Brazil	1,600	2,300★★	3,900
Chile	1,140★★	60	1,200
Argentina	460	180	640
Venezuela	180	20	200
Mexico	60	20	80
Other Latin America	80	350	430
Spain★★★	-	450	450
Portugal★★★	-	400	400
South Africa	500★★	800★★	1,300
Angola	20	50	70
Congo	-	40	40
Kenya	160	10	170
Zimbabwe	70	10	80
Other Africa	55	330	880
New Zealand	1,180	20	1,200
Australia	900	60	960
Other Oceania	50	30	80
Indonesia	★★★★	100	100
China	-	400	400
Other Asia	-	170	170
WORLD	18,950	6,300	25,250

★Borderline fast-growing with 45-year rotations, largely for sawnwood. ★★ Very fast-growing, largely for paper pulp. ★★★Borderline fast-growing, pulp-oriented. Iberia also has 4m ha of slow-growing conifers. ★★★★Indonesia also has about 0.7m ha of slow-growing conifers.

Source: Bazett 1993.

Much native forest in the US, Japan and Europe was cleared long ago and replaced by agriculture, pasture or tree plantations (Westoby 1989). While industrial wood remains physically abundant in all three regions, access to that wood is often blocked for economic or political reasons. In the South, meanwhile, forests are disappearing at a dizzying rate due to complex causes in which Northern imperialism is deeply implicated (Myers 1989, Colchester and Lohmann 1993, Vandermeer and Perfecto 1995). For dominant Northern interests and their allies among Southern elites, the threats created by deforestation are threefold: shortages of industrial wood needed to support indefinitely-increasing rates of per capita consumption; loss of other resources, including genetic materials, entailed by tropical deforestation; and loss of climatic stability due to rising levels of carbon dioxide.

Large-scale monoculture plantations are one way of responding to these crises of the prevailing economic model without addressing their underlying causes. They promise, for example, a way of meeting largely-Northern timber demand while putting off the question of how to stabilize or reduce that demand. Hence such statements as 'the predicted gap between wood demand and supply in many regions is so huge that only plantations will fill it quickly enough' (Sargent and Bass 1992). Such views are descendants of others prevalent in early 19th-century England, when depletion of domestic oak forests led to teak logging and teak plantations in Asia.

Industrial plantation programmes promise, as well, in the words of a study done for Shell International, to 'counter the greenhouse effect, either by serving as carbon sinks, or by alleviating pressure on native forests, helping to preserve them as carbon depots' (Shell/WWF 1993). As chapter 6 indicates, this promise is false in both particulars. Nevertheless, it has enough superficial plausibility to distract uninformed audiences from the more interesting topic of how to find alternatives to a system whose logic dictates a never-ending spiral in which ever-greater carbon emissions necessitate an ever more desperate search for carbon sinks.

Accompanying the rise of industrial plantations — and of radically simplified forests — has been the development of a modern forestry science which tends to add to their credibility. This science, silviculture, arose in the North mainly as a consequence of industrial development, with its huge needs for timber and agricultural raw materials; deforestation forced industrializing countries fairly early on into finding new ways of managing forests and reestablishing woodlands.

The new forestry science separated forest management strictly from agriculture, and focused almost exclusively on production of uniform quantities and qualities of timber. The multiple functions of native forests and diverse community woodlands were reconceptualized as symptoms of 'untidiness', 'disorder', and 'weediness'. Non-wood uses of forests were recast as 'minor products'. Trees whose growth rates had ceased in economic terms to justify their continued existence were dismissed as 'overmature'. Flora and fauna which lacked market value, or which reduced timber output, were classified as 'unproductive' and became candidates for eradication. Forests were to be replaced by a factory-like 'order' of stands of single varieties of commercially-valuable trees, of which plantations became the best examples. Seeds, plants, nutrients, growth rates and dates of harvest all became candidates for human control. Negative social and environmental consequences were played down as problems which could be 'mitigated'.

The problems modern forestry science sets and solves, in short, are those thrown up by a politics of centralized control of land aimed at extracting a very few types of raw material in industrial quantities. Working exclusively within mainstream forestry science means not asking questions about, and thus tacitly supporting, that politics. Forestry science is thus not a 'neutral tool' which can be detached from its social surroundings and adapted to any political purpose. It comes complete with a strong set of political biases, no less when it is practiced in a professional manner than when it is not. To appeal to it alone for an answer to the question of whether any particular industrial plantation is an appropriate use of land is therefore to guarantee social as well as intellectual conflict.

About this book

This book, too, is not a 'neutral tool'. (No such tools exist.) Unlike most forestry science and mainstream economics, however, it strives to be self-aware and straightforward about its origins, orientation, and audience. Growing out of a widespread concern over the expansion of pulpwood plantations in the South, it attempts to organize the information and analyses it draws on in a way which will be of practical interest to those who are alarmed at plantations' deleterious and anti-democratic effects and who are seeking alternatives.

Thus chapter 2, which sketches the political dynamics associated with a particularly large-scale, capital-intensive, cyclic industry, attempts to shed light on several questions often suppressed in mainstream discussions of pulp and paper. For example, why did destructive, extensive, highly-centralized plantations arise in the first place? Why do these plantations grow wood? And what are some of the dynamics lying behind rocketing per capita paper demand? In attempting to account for the recent shift toward monoculture plantations in the South, chapter 3 goes on to look at the pulp and paper industry's tendency toward increased centralization and globalization. The social and environmental consequences of large-scale industrial plantations are then spelled out in chapter 4.

Chapters 5 and 6 move to the topic of how, in practical political terms, these plantations are becoming a reality. Describing the different players who promote plantations (Northern and Southern governments, multilateral and bilateral development agencies, Southern elites, transnational paper corporations, banks, machinery suppliers, consultancies, academics, and even NGOs), chapter 5 probes the mechanisms by which, working in loose alliance, they reorganize landscapes, capture subsidies, and otherwise reshape social environments in ways friendly to themselves. Chapter 6 is devoted to investigating another aspect of the industry's expansion often neglected in economistically-oriented studies of corporate strategy: techniques for managing opposition.

The case studies of Part Two concentrate on the Southern countries which, so far, have had the most prominent roles in the international trade in plantation wood fibre — Brazil (chapter 7), Chile (chapter 8), South Africa (chapter 10) and Indonesia (chapter 11). For contrast, the experience of two other countries with plantations is also examined. These are Uruguay (chapter 9), where plantations degrade not forest, but sparsely-populated grasslands, and from which, astonishingly, plantation wood is being exported to the naturally heavily-forested Nordic region; and Thailand (chapter 12), where popular rural resistance to plantations established in thickly-inhabited areas has been persistent and often effective in challenging the spread of industrial plantations.

Chapter 13, finally, sketches a few of the many constructive ways forward being adopted by movements concerned with the spread of plantations. These include coordinated critiques of the enormous public handouts which make the plantation boom possible and which help it further redistribute wealth and security from poor to rich in an increasingly skewed world system; exposure of the political maneuverings of various industry alliances; serious approaches to the issue of demand creation; and promotion of alternative-fibre use in decentralized, less energy-intensive systems.

Looking beyond this book

It is important to emphasize, in closing, that the handful of wood-fibre exporters to which this book confines its examination are not the only, or even necessarily the most severely-affected, of the Southern countries undergoing a plantation boom. A more complete survey would have to examine the experience of many other nations.

India, for example, although not a significant exporter, has had an exceptionally long, extensive and painful encounter with pulpwood plantations. Following the nation's independence in 1947, the price of raw materials for paper was so heavily subsidized that the industry's profitability remained high even as selectively-logged forest stocks were depleted. In the 1950s, some mills were provided with bamboo at a cost of one rupee per tonne when the prevailing market price was over Rs 2000 per tonne. The explosive growth in capacity which this support enabled made the paper and rayon industry even hungrier for raw materials. Prodded by the United Nations Food and Agriculture Organization (FAO), India's forest department consequently authorized clearfelling of large tracts of natural forest which, with the help of more subsidies, were then replanted with eucalyptus, pine, and other monocultures. A wider range of species and ages of native trees over greater and greater geographical areas also began to be put to the blade, as previously-protected watershed areas were redesignated as selective-cutting and then as clearfelling/plantation areas.

Biological and social problems resulted which are now familiar in countries across the world. In Karnataka and Kerala, harvests from so-called fast-growing plantations were only 10–43 per cent of forest department predictions. Eucalyptus plantations in an area of the Western Ghats on which nearly 40,000 hectares of tropical evergreen forest had been clearfelled were wiped out when the *Cortecium salmonicolor* fungus took advantage of the concentration of so much of a single tree variety in a single high-rainfall location. Protests proliferated among villagers who were deprived of livelihood when the forests they had used for fodder, fuel and food were replaced with stands of commercially 'desirable' species.

Under misnamed 'social forestry' programmes bankrolled with state and foreign funds, trees were then planted on common lands, open government lands, and excess lands belonging to big and absentee landlords, mainly to the benefit of the pulp industry and other non-local actors. Millions of private farmers were also given free seedlings, technical help and soft loans to get them to plant eucalyptus for industry on their own land. In states such as Karnataka, Gujarat and Haryana, perhaps a million hectares of farmland were brought under eucalyptus. While the tree grew far better there than it had in the earlier plantations, and led to increased involvement on the part of 'social forestry' programmes in private farm forestry, it usurped land which would otherwise have been used for locally useful food crops such as ragi and cash crops such as cotton. When the government opened the country to cheap imports of wood, moreover, eucalyptus's reputation as a profit-earner suffered, and pulpwood-planting became less popular among private tree farmers after 1986 (*DTE* [Delhi] 31.8.1995). In the last decade, pulp firms such as Karnataka Pulpwoods, Ltd. have taken control of village common lands for plantations, igniting protests from commoners who point to violations of customary rights to graze cattle and gather wood and food. Increased intra-village and inter-class conflicts have been another result. A recent scheme to lease out 2.5 million hectares of forest land to the industry at subsidized rates has stirred further controversy (Guha 1988; Gadgil and Guha 1992; SPS 1989; Shiva 1991b; Shiva and Bandyopadhyay 1987; Shiva, Sharatchandr? and Bandyopadhyay 1982; Saxena 1992a, 1992b; *DTE* [Delhi] 31.8.1995).

China has also brought vast areas under plantations usable for pulpwood. There, the government has formulated stupendously ambitious plans to increase national tree cover from 12 to 20 per cent, while foreign investors hoping to stoke a potentially gigantic demand appear eager, with the help of the World Bank, to transform the country's current rather decentralized, small-scale, agriculture-based pulp industry into a more centralized, wood-based one. In Viet Nam, another big new target for international pulpwood investment, plantations already cover, on some estimates, between 1.5 and 6 per cent of the nation's land area. In the Philippines, plantations are used as a means of colonizing indigenous groups as well as supplying a pulp and paper industry chronically short of raw materials.

Other Southern countries where plantations are likely to lead to increasing destruction include Argentina, which in 1994 boasted 770,000 hectares of eucalyptus and pine plantations. The plantations have enjoyed state support since 1940, but have only recently become a source of export goods (mostly logs and sawnwood for pallets). As is the case elsewhere, the FAO has been a big plantation proponent, and further support is now flowing from the World Bank, which has recently approved a forestry development loan, and the European Community, particularly Italy and Spain. In 1991, moreover, the Italian government proposed a 'Green Development Plan' for Argentina — fortunately, it was never implemented — which was aimed at the plantation of six million hectares of fast-growing trees (mostly eucalyptus and pines), to be planted over 15 years to fill Italy's needs for wood.

Elsewhere in Latin America, US pulp and paper companies such as Simpson Investment and Stone Container are setting up pulpwood plantations in Mexico, Costa Rica and Venezuela to ensure raw material supplies (Paré 1992, Cheney 1992, Swann 1992), much to the concern of a growing environmentalist public. In the Dominican Republic, a public battle had to be fought by conservationists in 1990 to stop eucalyptus from being planted in watersheds (Thomen 1990).

In the Congo, meanwhile, Shell has joined the Unité d'Afforestation Industriel du Congo in a venture called Congolaise de Développement Forestier. This firm owns 40,000 hectares of very fast-growing clonal euclayptus plantations at Pointe Noire, the wood from which, harvested every seven years, is sent to pulp mills in the North. While, according to Shell, the plantations 'are established as a mosaic fitting into the natural landscape' (Shell/WWF 1993), the firm's own environmental management plan notes that they are changing the open savanna into a 'veritable eucalyptus forest' (Geerling, N'Sosso and Kitemo 1991).

Plantations aimed at export production are also expanding in many warmer regions of the North, for many of the same reasons that lie behind their growth in the South, including quick yields and cheap land. Social and environmental impacts are also similar, as are, often, styles of resistance.

In Iberia, a growing fibre-export region, plantations have moved onto both commons and private land, just as they have in India and Thailand, eroding customary forms of livelihood and security and widening class gaps. Between 1940 and 1983, over 273,000 hectares of eucalyptus and 2,668,000 hectares of pine were planted in Spain, coinciding with a decline in rural economies and migration from rural areas. Yields have often been lower than expected, local employment has suffered, water cycles have been disrupted and fires have increased sharply. As in India, the trees have tended to help non-local powers such as absentee landowners make profits out of poor soils without having to be dependent on local communities. They have also degraded soils and wildlife refuges and replaced landscapes capable of sustained yields of products such as game, livestock, honey, herbs, firewood, cork, carving wood, and sheep (Bermejo 1995). In both Spain and Portugal, many of the same forms of protest are seen as in Asia, including the ripping out of seedlings and saplings by angry local villagers.

Nor do South–North parallels stop there. In Australia, as in Indonesia and Papua New Guinea, both natural forest and plantations have been exploited for pulpwood for export, resulting in public outrage. In New Zealand, as in Chile, large temperate areas are being planted with pine. Even in the southeastern United States, an old wood-producing area where more and more forests are being chopped down for export and converted to pine plantations, environmental opposition is having a strong impact. In Nordic nations, finally, environmentalists are contesting the pulp and paper sector's approach to forest lands at home in much the same terms Southerners use to criticize the plantation model exported by Nordic consultants (Olsson 1995).

This book cannot tackle any of these topics in detail. It is hoped, however, that it will prove a useful resource not only to Southerners threatened by the advance of export pulpwood plantations, but also to others concerned with forest industries and their globalization generally.

Chapter 2

Introducing Pulp and Paper

The evolution of a wood-based industry

Paper consists of a web or mat of cellulose-based vegetable fibres which have been refined and treated in water before being deposited on a screen and dried. First made in China around 2,000 years ago from vegetable waste materials, paper reached India by the 600s, Turkestan by about 750, Damascus and Egypt by the 800s, Spain by the 1100s, and Northern Europe by the 1400s. The first raw materials included silk, hemp, cotton rags, old fish nets, mulberry bark, grasses, bamboo, flax, laurel and rattan. Early European papers were made mainly from linen or cotton rags (Grant 1978, Western 1979, Wilson 1991).

With the advent of printing, paper use increased, but production remained limited by available technology. Demand, too, was constrained by the virtual absence of a paper packaging industry or a mass newspaper market. In the 19th century, however, French and English factory owners, struggling to overcome the power paper artisans held by virtue of their specialized knowledge, began to develop, with the help of the industrial revolution's new machine tool industries, paper machines which centralized paper-making technique in capitalist hands. Fitted with endless wire meshes which rotated like conveyor belts, modern paper machines began to appear around 1800 (Hills 1988, Hunter 1978, Clapperton 1967, Coleman 1958).

Prevailing non-wood raw materials were not ideally suited to the large, centralized plants such machines eventually made possible. Rags were too scarce and expensive to keep large mills running at capacity. Straw, which began to be widely used in agricultural countries such as Germany and France, and esparto grass, which was imported from Spain and North Africa into Britain in large quantities from 1880 up until the Second World War, were available in greater volume. These materials, however, had drawbacks. For one thing, they could be supplied only seasonally, were vulnerable to crop failures and, if they were to be supplied in large quantities, required extensive catchment areas. They were thus far from perfect raw materials for big, capital-intensive mills which had to be run year-round. Agricultural raw materials were also bulky and had to be baled in the field if they were to be transported to distant mills. It was not until the 1860s, when pulps made from wood finally began to become commercially feasible, particularly in North America and the Nordic countries, where old-growth and other forests provided relatively compact, contiguous, large-scale 'mines' of raw material, that the modern Western paper industry really took off. The price of newsprint and other types of paper, measured in conventional economic terms, dropped by about 85 per cent. Demand was spurred, and the newspaper and other paper-dependent industries grew quickly. Today paper production represents one per cent of the world's total economic output (Ionides 1994, *Nation* 22.2.1995, Chaudhuri 1995, Grant 1978).

The use of wood has only reinforced reliance on large, highly-mechanized mills. Chipping equipment and stone grinders used to process logs, for one thing, produce too much for small mills to absorb. The more pulp-mill technology becomes geared to wood, in addition, the less inclined industry becomes to adapt to other raw materials, preferring to seek alternative wood sources in times of crisis. Today, wood is used for around 90 per cent of world paper pulp production, which comes to over 170 million tonnes per year. Some 640 million cubic metres is consumed annually in the process, or nearly 13 per cent of total world wood use, the rough equivalent of the mature timber that would cover over two million hectares, an area half the size of Switzerland. Each issue of a mass circulation daily in Britain would consume about 10,000 trees if its newsprint contained no recycled fibre; the lifetime newspaper reading of an average citizen in the US or Japan requires the pulp equivalent of several hundred trees, most of which go toward advertisements. In order to maintain such levels of consumption without cutting fully-grown natural forests, about ten million hectares of land worldwide would have to be planted with trees each year (Ayres 1993, *PPI* 10.1994, Shell/WWF 1993, Judt 1994, Grant 1978, Gauthier 1991, IIED 1995, Durning and Ayres 1994, Wright 1995).

The first woods used for paper were poplar and willow — largely because they then had few other commercial uses. Interest soon shifted to softwoods such as pine, spruce and fir, however, since their cellulose fibres were longer (two to five millimetres) and produced stronger, higher-quality pulps. In 1987, some 46 per cent of pulpwood consumption was derived from coniferous roundwood, 27 per cent from coniferous industrial residues, 22 per cent from hardwood roundwood, and five per cent from hardwood residues (Cardellichio et al. 1989).

Following the Second World War, Japan revived and advanced commercial technology for using hardwood trees, such as birch, beech, alder and mangrove species, which yield fibres only 0.5 to two millimetres long. This enabled the country's paper industry to exploit not only domestic broadleaved trees, but also Southeast Asian mangroves, Australian native eucalyptus, South African and US plantation hardwoods, and Chilean beeches and the mixed hardwood forests of Papua New Guinea. Australian improvements in technology for pulping eucalyptus assisted in the process of converting many of that country's native forests to paper, and added impetus to the spread of eucalyptus plantations in Asia, Latin America and Africa. As the industry shifts its quest for raw materials to the South, the proportion of hardwood to softwood pulps grows, with hardwood pulp now comprising over 40 per cent of total pulp trade. Eucalyptus in particular is increasingly attractive as a raw material for computer, copy, fax, high grade printing, tissue and other papers, and trade in eucalyptus pulp is increasing considerably faster than that in other varieties (JATAN 1993, Wright 1993, Marchak 1992, Schreuder 1988, Kroesa 1990).

The reliance on wood and on large mills which is characteristic of the industry's mainstream today, however, is the result of historical momentum, not scientific or economic necessity. Even the exorbitant rates of consumption in the West and parts of East Asia do not necessarily entail a tree-based paper economy. In some countries, non-wood raw materials such as straw, bagasse (from sugar cane), bamboo, cotton linters, sisal, seaweed, abaca, reeds, esparto and other grasses predominate. Some 60-65 per cent of China's paper, for example, is produced using straw, bagasse, cotton waste, and other vegetable fibres, while esparto grass provides the raw material for Tunisia's domestic pulp-making industry. Many observers also see the proportion of India's paper produced from agricultural wastes, now between 30 and 45 per cent, rising in the future (Bayliss 1995). In 1991, 32 per cent of the pulp used in the South was derived from non-wood materials (Dudley, Stolton and Jeanrenaud 1995). Although the proportion of non-wood-based paper is far smaller in the North, there is no reason why this state of affairs should hold indefinitely. Indeed, evidence is now emerging that hemp would be more widely used as a raw material in the US today had not wood-fibre businesses made an astute political alliance earlier in this century with institutions interested in promoting a drug panic over hemp cultivation (Hanson 1995).

According to some observers, the ratio of non-wood to wood raw materials is increasing worldwide, with over 300 industrial-size mills now using non-wood fibres (Paavilainen 1993). In the view of researcher Maureen Smith, there are no purely technical obstacles even to the US's exorbitant current paper demand being met entirely by a decentralized network of small- to medium-sized mills using regionally-appropriate non-wood raw materials (Smith 1995, Smith forthcoming). Year by year, hemp or kenaf (a jute-like crop) produce more fibre of good quality than wood, and using agricultural and other wastes as raw material is generally not only more efficient, but also more socially productive and environmentally beneficial than using wood (Ayres 1993; Western 1979; Wright 1994; Bayliss 1995; Riddlestone et al. 1995; *DTE* [Delhi] 31.8.1995).

From wood to pulp

Once the water is taken away, a tree trunk is only about 50 per cent cellulose. The rest consists of about 30 per cent lignin (a tough, resinous adhesive that provides structural support to the tree) and 20 per cent oils and other substances. Making a tree's cellulose available in a form which can be used to make paper can only be done either by grinding up the wood (to make mechanical pulp) or by chipping it and boiling the chips with chemicals before refining (to make chemical pulp).

Mechanical processes turn up to 95 per cent of the wood into pulp but tear the fibres, shortening them and weakening the resulting pulp. Mechanically-pulped fibres can thus be recycled only three to four times, as opposed to chemically-pulped fibres, which can be used five to ten times. Mechanical processes also leave lignin in the paper, which causes it to turn yellow when exposed to light. As a result, mechanically-produced paper is used mainly for newsprint, telephone books and other products where strength and quality is not at a premium.

Chemical processes — which account for more than 75 per cent of world production — produce a stronger pulp because they do not damage the wood fibres. In addition, the most important chemical processes used separate the lignin from the cellulose, making possible the production of papers that do not yellow with age. (Although the industry calls these 'wood-free' papers, this term does not mean that they are not made with wood. A more proper term would be 'lignin-free'.) As a result, only between 45 and 65 per cent of the wood is turned into pulp. Producing one tonne of bleached chemical pulp from fresh raw material requires 120,000 or more litres of water, over 20 plantation trees or 4.8 cubic metres of wood, and approximately 1.2 megawatt-hours of electricity. Overall, this is as much energy as is required for the production of a tonne of steel. Wood residues are today typically burned to boil chips and generate electricity and steam for the pulping process. Thus although modern chemical mills need more water, electricity and heat than mechanical ones, they tend to require less energy from outside sources such as thermal plants or dams (Grant 1978; Oinn 1994; Smith forthcoming; IIED 1995; Floegel

1994; Kroesa 1990; Dudley et al. 1995). In terms of energy and water use per unit of paper manufactured, currently-dominant chemical and mechanical wood-based pulping processes are without question both far less efficient and far less sustainable than traditional methods.

There are several ways of making chemical pulp:

- The sulphate, or kraft process, which involves boiling wood chips with caustic soda, produces 95 per cent of the pulp traded on the open market. This process produces a strong pulp which, although dark brown at first, remains white long after it is bleached due to its low lignin content. At least 95 per cent of the chemicals used in the kraft process are recovered and reused, but between one to three kilogrammes of sulphur dioxide are released to the air for each tonne of pulp produced, with potential effects on soil, water, and the health of humans and plants.

- The sulphite process boils wood chips in an acid solution, yielding a light brown, strong, soft pulp. The sulphite process also reuses chemicals, but emits more, around five kilogrammes of sulphur dioxide per tonne of pulp, and the damage caused over the past century by the water pollution associated with this process is inestimable. As with the sulphate process, cellulose fibres lost during processing are discharged into waste water, where they decompose, depleting the oxygen dissolved in the water.

- The chemo-thermi-mechanical process vapour-heats and chemically pre-treats wood chips to remove some lignin and resin, then grinds them up to produce a fairly strong, soft, slightly yellow pulp often used to make tissues and some writing-grade and coated papers. This process can be used for both softwood and hardwood, and usually discharges not only the wood chemicals removed from the pulp, but also the sulphur added in the pulping process, creating a highly-toxic, persistent effluent.

Papers produced by either mechanical or chemical processes require bleaching. Yellow mechanical pulps are usually bleached with hydrogen peroxide, while dark brown kraft pulp requires heavier bleaching, traditionally with chlorine or chlorine dioxide, but now, increasingly — as a result of environmentalist campaigns and consumer pressure — with oxygen, ozone or hydrogen peroxide. Chlorine and chlorine dioxide, while they are effective in removing lignin and in strengthening pulp, react with organic chemicals present in pulp to form hundreds of organochlorine pollutants including dioxins, which are some of the most potent poisons known. (See chapter 4.)

From pulp to paper

Most pulp is produced in integrated pulp and paper mills and goes directly into paper manufacture. Approximately 17 per cent, however, is dried and traded internationally to non-integrated paper mills, sometimes at a great distance. This ratio has increased only slightly since 1980, when the figure stood at 16 per cent (IIED 1995), and the trend among the largest paper manufacturers is to reduce dependence on pulp bought from outside the company (Higham 1995). In 1993, the South produced less than one-fifth of the world's total paper pulp output of 167 million tonnes and just over one-fifth of its total paper output of 254 million tonnes (see Tables 2.1 and 2.2) (FAO 1995).

TABLE 2.1
World's top pulp producers, 1994

Country	Pulp production (million tonnes)	Per cent of world production
US	58.7	34
Canada	24.5	14
China	17.1	10
Sweden	10.9	6
Japan	10.6	6
Finland	10.0	6
Brazil	6.1	4
CIS*	3.3	2
France	2.8	2

*Former USSR minus the Baltic states.

Source: PPI 7.1995

To make paper, different types of wet pulp are blended, mixed with fillers (calcium carbonate, kaolin, titanium dioxide, and so on) and other additives (rosin, aluminium sulphate, dyes), spread into an even sheet on a wire mesh, dried, then removed with an absorbent felt. The surface of printing and writing papers is then smoothed mechanically or coated with clay or chalk. Out of at least 34 different categories of pulp, over 420 commercial grades of paper can be produced, with individual mills often being capable of producing a variety of papers from the same forest or plantation (Fernandez Carro and Wilson 1992).

TABLE 2.2
World's top paper producers, 1994

Country	Paper production (million tonnes)	Per cent of world production
US	80.7	30
Japan	28.5	11
China	21.4	8
Canada	18.3	7
Germany	14.5	5
Finland	10.9	4
Sweden	9.4	3
France	8.7	3
Italy	6.7	2
South Korea	6.3	2
Brazil	5.7	2
UK	5.5	2
CIS	4.8	2
Taiwan	4.2	2

Source: PPI 7.1995

Newsprint (the paper used for newspapers), which typically has a weight of 40–49 grammes per square metre (gsm), is made mainly of mechanical (lignin-containing) pulp with few or no fillers added. Fine paper (printing, writing, computer and business communication papers), which tends to be thicker and heavier, is almost always made of lignin-free and highly-bleached chemical pulp, although sometimes chemi-thermi-mechanical pulp is also used.

Printing and writing papers range from 50 to 350 gsm in weight. Coated paper, used in commercial printing and glossy magazines (largely for colour advertising), as well as illustrated books, bears a surface layer of pigment particles finer than the fibres which make up the paper itself. Either chemical or mechanical pulps can be used, although the former currently predominate. Stationery and photocopy papers are uncoated, weighing in at around 70–120 gsm.

Sanitary products such as toilet paper, tissues, napkins, and sanitary towels require that resin acids and other natural wood chemicals which prevent wood fibres from absorbing water be removed. Sulphite pulp made from softwood has been the preferred raw material for most tissue and towels because of its softness. For stronger towels, kraft pulp may be used. For toilet and other absorbent paper, recycled fibre can be used. Fluff pulp, used together with added materials for absorbence to make disposable diapers, is made from sulphate or chemo-thermi-mechanical pulp. Brown wrapping or bag papers are made from softwood kraft pulp, bleached or unbleached.

Card, with a weight of 160 gsm and above, and board, with a weight of 220 gsm and above, are generally used for packaging. Linerboard is made from unbleached kraft softwood pulp. The corrugating medium which forms the 'filling' between two sheets of linerboard in box-making material is made from unbleached, chemi-thermo-mechanical pulp, usually made of hardwood, as well as recycled fibre. Board may incorporate unbleached or bleached kraft pulp for strength, with many containers being coated with waxes or plastics. Some corrugated packaging is made from a sulphite process which involves no bleaching.

Cellulose films and rayon are made by a modified kraft or sulphite process which uses intense chlorine bleaching to remove all lignin and wood ingredients. The pulp is further treated chemically, regenerated in sulphuric acid, and then forced through holes to produce rayon or through slots to produce cellophane. Self-copying papers contain ink in small droplets of wax or solvent, while fax paper contains a layer of heat sensitive pigments (Kroesa 1990, Paper Publications 1994, Oinn 1994, Biermann 1993).

From paper back to pulp

For over a century waste paper has been used for making new paper wherever it has been economic to do so. What with environmentalist pressures, recycled paper is today more important than ever as a raw material, being used increasingly in newsprint, writing papers, and toilet and tissue papers. Behind this growing use of recycled paper are both technical advances which improve the quality of recycled paper and consumer groups who point out that for most uses, paper need not be of a standard which requires a high ratio of fresh fibre.

With the globalization of the pulp and paper economy, waste paper has also become an important item of trade. Some 16 per cent of world waste paper consumption entered international trade in 1992, with huge quantities exported within Europe and from the wood-rich, overconsuming US to wood-poor economies in Asia. When woodpulp prices are high and recycled paper in great demand, as in the early 1990s, waste paper becomes an even more attractive raw material.

Recycled paper can be made from either pre-consumer waste or post-consumer waste. Pre-consumer waste consists of unprinted industrial by-products such as printer's trim or paper mill waste; post-consumer waste includes already-printed paper and used corrugated cardboard from offices, newspapers, shippers and homes.

TABLE 2.3
Waste paper use and collection, selected countries, 1992

Country	Use rate* (per cent)	Recovery rate** (per cent)
Taiwan	98	56
Denmark	97	37
Mexico	81	35
Thailand	80	35
South Korea	70	43
Netherlands	70	53
UK	60	32
Japan	53	52
Indonesia	53	17
Germany	52	51
US	33	39
China	32	24
Brazil	31	37
Canada	17	35
Sweden	14	44
Finland	5	28

*Ratio of wastepaper used in production to amount of paper produced.
**Ratio of waste paper recovery to paper consumption.
Sources: P&P 10.1993 (quoted in Ryan 1994), IIED 1995, FAO 1994.

Although the amount of recycled paper used varies widely from country to country, waste paper accounted for approximately 18–20 per cent of the material input for the world's paper output in 1970. With environmental pressures, this figure had risen to around 30–32 per cent by 1988 and perhaps 35–37 per cent by 1995. Top consuming countries such as Taiwan, Japan, Germany and The Netherlands are major users, but Southern countries for whom waste is easier to obtain than wood generally use a greater proportion of it in their raw material supply. About half of the raw material for Asian-made paper consists of waste, and nearly 45 per cent of the raw material for Latin American-made paper, but only about 28 per cent for North American paper and 37 per cent for European. Overall, the South uses a raw material which consists of approximately 10-15 per cent more waste paper than does the North (IIED 1995). Many Northern countries, on the other hand, are more assiduous waste paper collectors. However, they often wind up dumping waste on the international market rather than using it themselves (see Table 2.3).

Industry sources suggest that the global use rate for waste paper may reach 42-45 per cent by the year 2000, while FAO suggests it is more likely to stagnate (Niku 1993, IIED 1995). At present, the use of recovered fibres in paper-making is growing twice as fast as paper production itself. This, of course, cuts into the demand for wood pulp. A mere one per cent increase in the use of waste fibres in paper raw material in Finland alone, for instance, would save 376,000 cubic metres of wood per year (FAO 1994). Roger Olsson of Taiga Rescue Network estimates that a world recycling rate of 50 per cent in the year 2010 would save 200-300 million cubic metres of roundwood annually at projected consumption rates. This would obviate the construction of around two giant woodpulp mills per year between 1990 and 2010 (Olsson 1995). If the US had adopted the over 50 per cent recycling rate of the Netherlands in 1987, 500,000 fewer hectares of forests would have had to have been logged per year for conversion into paper pulp (Graham 1994). Also reducing the rate of rise in demand for pulpwood, although to a lesser extent, is the increasing use of mineral fillers and coatings in paper. The proportion of pulp in paper-making furnish is estimated to have dropped from 65 to 64 per cent between 1993 and 1994 alone (Ayres 1993, Clark 1994, Shell/WWF 1993, McClelland 1994, Niku 1993, Brennan and Pappens 1995).

Newspapers can be de-inked easily, making the reuse of newsprint attractive. Cardboard, too, can easily be recycled, and printing and other pre-consumer waste, office bond, and discarded photocopies can be converted into new stationery and copying paper if kept separate. Mixtures of all grades can be recycled into low quality products such as egg cartons. De-inking photocopies and laser-printed material, however, requires newer technologies. In addition, highly coated paper, envelopes with windows and paper containing synthetic glue (self-adhesive envelopes) are hard to recycle; and fax paper, carbonless copy paper and plasticized drink cartons cannot be recycled at present. It is difficult, moreover, to make extremely white paper out of recycled fibres without using harsh bleaches.

Like all 'technical fixes', recycling cannot in itself be equated with wood conservation, but must be viewed in political and economic context, alongside an analysis of demand, trade and industry structure. Even if the world recycling rate increases to 50 per cent, for example, a 12-15 per cent increase in the world's industrial roundwood production will be required by 2010 if FAO projections of consumption increases are borne out. In the US, the increased use of recycled pulps in paper production has not even *slowed* the rate of growth in fresh woodpulp production, instead merely increasing woodpulp exports (Harland 1994, Olsson 1995, Smith forthcoming). Unless combined with institutional restructuring, moreover, more use of waste paper, instead of reducing dependence on industrial tree plantations, may merely spur the wood industry into attempting to create demand for alternative plantation products. For example, the FAO European Forestry Commission and the UN-ECE Timber Committee concluded in 1993 that because of the weakening demand for small-sized wood which has resulted from increased recycling, 'new outlets for small-sized wood should be developed' in energy production or other fields (FAO 1994). The Brazilian firm Aracruz is meanwhile hoping to create new markets for its plantation eucalyptus in construction, furniture, fibreboard and plywood in order to reduce commercial vulnerability to pulp price swings (*FT* 21.6.1995).

Moreover, although making paper from recycled fibre, as from vegetable fibres, tends to use less water and energy than producing it from wood, and to result in less pollution, the expenditure of water and power can still be high (Smith forthcoming; Dudley et al. 1995). In addition, both the proportion and the grades of paper that can be made from recycled material are limited. Fibres become shorter and weaker in the recycling process, making them less useful for papers requiring strength. Even the strongest, longest fibres turn into useless dust if they are recycled more than ten times, and most wood fibres have a far shorter life span. Pulp made from waste paper must thus often be topped up with longer-fibred pulp made directly from trees, hemp, kenaf or other materials in order to ensure strength. Although it is theoretically possible to restrict the overall ratio of fresh to recycled fibres to little more than 20 per cent (Dudley et al. 1995), a totally cyclical paper economy is not on the horizon. Mechanically-pulped recycled fibres, moreover, cannot be used to make certain quality grades of recycled paper which require chemically-pulped fibres.

Recycling requires, moreover, that ink, fillers, coating materials and staples be removed. The recovered ink, which is likely to contain barium, copper and heavy metals, tends to be discarded together with unrecoverable paper fibres, which may contain poisonous dioxins and furans. It is either incinerated, adding these pollutants to the air, landfilled, or spread on farms and gardens (Durning and Ayres 1994).

The dangerous effects of such discarded compounds underline the importance of pressing for use of nontoxic inks as part of campaigns to check overconsumption, reconsider waste disposal policies and use a greater proportion of recycled fibre in paper manufacture.

Large scale, capital intensity and centralization

As a growing pulp and paper industry using local wastes transformed itself, during the 19th century, into a highly-mechanized, centralized one requiring the destruction of large swathes of forest, literacy and the demand for cheap paper grew. Many of the new paper mills built around the beginning of the 20th century were designed to produce newsprint. These were of standard size and relatively inexpensive and profitable. By the 1930s, however, prestige competition among North Atlantic newspaper corporations, conjoined with technological advances following on from the First World War, had given an incentive to machine manufacturers to design larger and larger machines. Many of these turned out to be one-offs. Whereas in the 1900s, new newsprint machines tended to be 2.25 to 2.5 metres wide, running at speeds from 100-150 metres per minute, by 1937 machines were being built up to 7.7 metres wide, running at 420 metres per minute. The kraft sack and wrapping paper industry, which grew swiftly from 1930 onwards, soon became another source of orders for big machines.

Such machines became less and less cost-effective. Not only were many machines of a unique design; huge widths and speeds also required sophisticated and expensive controls for efficient operation, adding yet more to costs. At the same time, paper prices could be increased only so much, since they were set by the mass of earlier, smaller, less sophisticated, cheaper machines still in operation. By the 1960s, moreover, when widths of nine metres and speeds of 700 metres a minute were achieved, capacity began to exceed consumption by a wide margin and paper prices slumped. The cost per annual tonne of a newsprint machine increased at least 40-fold between 1930 and 1975, while the price of newsprint increased less than 20-fold. Yet by this time, machine manufacturers' investments in large machine tools had made it difficult for them to produce for anyone but the biggest paper companies. As paper expert A. W. Western (1979) notes, building new paper machines

> became a luxury which could be afforded only by multinational giants or the governments of developing countries, advised by consultants that only scale to this degree could be economic! For the consultants it *was* economic; they were now essential for large mill design and coordination.

Today a single new world-class pulp mill can cost as much as US$1 billion. In many countries only large manufacturers can afford even to renew their plants, and then only to adapt them to producing more specialized products with a greater profit margin.

The nearly 200-year-old dream of concentrating paper-making power in the hands of plant owners, in short, had been realized with a vengeance. Access to the dominant stream of paper-making knowledge was now restricted not just to capital, but to *big* capital. For any capital-short Southern society with an interest in meeting its own paper needs efficiently with indigenous materials, and in a way which did not require centralized control of large areas of land, the implications were particularly bleak.

They were also bleak, however, for the industrialized North. Partly because today's immense mills cannot generate profits without a large-scale re-engineering of their social and physical surroundings, the pulp and paper industry relies heavily nearly everywhere on political campaigns to capture handouts from the state and the public. As the giant Canadian firm MacMillan Bloedel instructed Alberta's Premier Harry Strom in 1969, when the company was seeking low-interest loans, tax exemptions, tax investment credit, and infrastructure subsidies from the province, 'pulp mills being built today are not profitable unless some special low cost conditions or concessions prevail' (cited in Pratt and Urquhart 1994).

Such subsidization, by making paper firms profitable, can in turn motivate them to invest in yet more and bigger machines and seek even bigger subsidies, with disastrous results for raw materials catchment areas. As historian Ramachandra Guha and ecologist Madhav Gadgil note, beginning in the 1950s, India's forest industries

> were subsidized so heavily, and could hike up the prices of their produce so freely in a seller's market, that their profitability has remained high, even as forest stocks have plummeted. . . . Even in the 1980s, bamboo prices were raised only to Rs200 to 500 per tonne, when market prices were well over Rs5,000 per tonne. The result of this state-subsidized profitability has been an explosive growth in industrial capacity, and a non-sustainable use of forest stocks (Gadgil and Guha 1992).

The networks through which a highly-mechanized and -centralized industry captures the subsidies it needs for survival will be explored in chapter 5 and in the case studies of Part Two.

While ownership of today's mainstream pulp and paper sector is not as concentrated as that of some other basic industries (the top ten paper businesses control only about a fifth of the international paper market, and no single market pulp producer has more than a six per cent market share), it is dominated physically by relatively few large plants. The US, the world's biggest producer, has only 203 pulp mills, implying an average mill capacity of over 300,000 tonnes per year. In Japan 49 pulp mills are capable of producing an annual average of nearly 310,000 tonnes each. Finland holds its position as a leading pulp and paper exporter with a mere 43 pulp mills averaging over 250,000 tonnes per year in capacity. The relatively young export pulp industry in the South also revolves around a small number of huge plants. In Chile six pulp mills boast an average capacity of 350,000 tonnes a year, in Brazil 35 are able to churn out an average of 175,000 tonnes each and Indonesia's 13 mills have an average annual capacity of 215,000 tonnes. China's huge pulp production, by contrast, is spread over about 8,000 mills whose average output is well under 2,000 tonnes per year. Some mills, like village bakeries, may be open only two days a week. Not coincidentally, such mills tend to rely on local, non-wood sources of raw material. In India, too, average pulp mill capacity is low: 13,500 tonnes per year (*PPI* 7.1995, Wright 1994).

Unsurprisingly, the large-scale mainstream industry generates very little employment per unit of economic output. In the US, for example, paper and pulp is the most capital-intensive of all manufacturing industries, and twice the industrial average, with more than US$120,000 of plant and equipment invested in every employee. The world-class mills being built in Indonesia, Brazil and elsewhere, which use the same type of equipment, are hardly less so. Approximately $750,000 in capital, for example, is being invested for every job created in the new pulp mill at Riau Andalan in Indonesia (see chapter 11). Similarly, $700,000 is being invested for each job in two new cellulose, paper and thermal energy complexes in Galicia, Spain being backed by the German multinational Feldmuhle Aktiengesellschaft and the Finnish firm Tampella Oy. The cost of each of the 365 jobs created at the enormous new Al-Pac bleached kraft pulp mill near Athabasca, Alberta, owned by a Japanese-Canadian consortium, is meanwhile a staggering $1.3 million. Employment, already low in the mainstream industry, is moreover set to decline even further. Employment in the US's paper industry, for instance went down one per cent between 1993 and 1994 alone, due mainly to mergers and the phasing out of older technology. Between 1990 and 1992, Canada's forest industry eliminated 62,600 jobs, some 28 per cent of the direct workforce. Knowledge and skills are another casualty of the industry's capital-intensivity. The dominance of gigantic machines ensures that the opportunity to learn about and use paper-making technology is restricted to a select few technicians and technocrats (Van Hook 1994, Western 1979, McClelland 1994, CEPA 1992, Olsson 1995, *TAPPI Journal* 1.1995, Smith forthcoming, Schindler 1995).

Boom and bust

Reliance on big, expensive, centralizing machines — combined with freely-available technology, easy availability of debt finance and wood, and little need for newcomers to buy into brand names — tempts the industry, whenever demand increases, into building huge amounts of new plant to supply it. Big firms, in addition, have sometimes hoped that being the first to build a giant new machine during boom times will scare off competitors and gain them enough market share to become price-setters. In what is still a crowded field, however, such hopes have proved vain. As a result of overspending, the market becomes glutted with pulp and paper a couple of years after the market peaks — it takes from 18 months to two years to bring a new pulp mill on line — sending prices into a deep slump. Left with enormous machines on its hands which cannot be run at full capacity, the industry finds it difficult to pay off its debts and ceases to make a profit (Wright 1993, 1994; *P&PA* 11.1993; *PPI* 9.1994).

Such cycles afflict pulp and paper even more severely than they do other basic industries such as chemicals or metals. They are possibly exacerbated by the behaviour of banks, which are the biggest suppliers of finance to forestry industries, and from which cheap funding is especially easily available at the peak of the paper and pulp cycle. Adding further to the industry's volatility is the growing tendency of firms to invest across international borders, which often gives exchange rate fluctuations a huge importance in determining a company's profitability (Fletcher 1988, van Dijk and Dekker 1995, Clark 1994). Government policies also do their bit. Changes in US tax laws in the 1980s, for instance, gave corporations tax breaks for debt instead of equity, giving firms extra incentive to take on debt by building new mills (Floegel 1994). All in all, paper and pulp's wild market swings appear difficult to control within the current system.

During the boom years of the 1980s, for example, when demand was surging and pulp and paper were selling at high prices, leading paper industrialist Hugh Fletcher of Fletcher Challenge warned his colleagues that although 'the industry does not have a history of being logical', this time it really should refrain from investing too much in large new plants. The advice went unheeded. By 1993, pulp prices, in constant dollars, were half of what they had been only four years previously, and 39 per cent of what they had been in 1975. By 1994, Ronald Y. Oberlander, President and Chief Executive Officer of Abitibi-Price, which had lost over US$460 million in the period 1990-4, was lamenting over the sector's 'inability to manage the periods of prosperity and our inability during these periods to spend money wisely'. Canada's pulp industry as a whole had lost more than $4 billion between 1991 and 1993. In December 1993, Avenor's new Gold River newsprint complex in Vancouver was forced to cease production, joining other plants such as Abitibi-Price's newsprint mill in Thunder Bay, Ontario. In Japan, excessive supply and world economic recession brought the market to a standstill, forcing companies such as Daishowa Paper into a serious management and debt crisis (Fletcher 1988, Oberlander 1994, Soulas 1994, McClelland 1994, Avenor 1993). Chile's industry meanwhile lost up to US$200 million in income per year; Latin American earnings suffered more than those of any other industry region in the early 1990s. At the same time, Indonesian pulp production dropped to a mere 65 per cent of capacity, and producers such as Thailand's Phoenix Pulp and Paper had to stop exporting.

Yet between mid-1993 and late 1995, Northern softwood bleached kraft pulp increased in price from US$390 per tonne to nearly $1,000. This allowed pulp mills to operate nearer capacity, but threatened the survival of some firms in the overcapacity-ridden paper industry, who were hard put to pass on such large price increases to consumers.

Among the reasons the industry cited for the spectacular reversal were economic recovery in the US and Europe; increased paper demand in Europe, Southeast Asia, and the US; the closure of some integrated pulp lines and an associated run on market pulp; a sudden hoarding response to pulp producers' announcements of price hikes in late 1993 among buyers who had confidently allowed their stocks to be depleted; a European hardwood shortage and a rise in pulp production costs there and in western Canada; weak US and Canadian dollars, which made North American pulp and paper exports cheaper, stimulating demand; rumours of a strike at pulp and paper mills in British Columbia; and the extreme depth of the previous slump, combined with new environmental requirements, both of which discouraged new investment which could have increased supply. Also significant was a tradition of noncooperation between pulp buyers and sellers: pulp dealers who feel they are being pushed into offering ever lower prices in a slump tend to get even by squeezing their customers as hard as possible when the market begins to improve. The same holds for paper dealers (Bingham 1995; *Independent* 19.6.1995; Pappens 1995; Edwards 1995; Stefan 1995; Rahikainen et al. 1995).

By 1996, though, the market was once again diving, with Northern softwood bleached kraft pulp dropping 50 per cent to US$500 per tonne in May before beginning to rebound. Frightened by the prospect of price increases, buyers had built up stocks which they began to draw on as the economy worsened, reducing orders. At the same time, as one trader complained, hardwood pulp suppliers from Brazil, Indonesia and Russia 'have lowered prices, but they have not created demand', with Chinese purchases being especially disappointing. As capacity increases, further drops may occur in the future (*FT* 23.11.1995, 5.1.1996, 24.1.1996, 8.2.1996, 2.5.1996, 8.5.1996). Some industry observers hope that, due to constraints on spending forced by environmental expenditure, future slumps will be less severe (Bingham 1995; *FT* 13.7.1995; *Economist* 14.1.1995). In the view of executive David Clark, however, 'the paper cycle shows no sign of diminishing. Indeed, the swings show every sign of increasing as the business becomes more global and more capital intensive with more aggressive competition among larger groups' (Clark 1996).

Such slumps, of course, can often bring advantages to large corporations with the resources to weather them. Firms such as Indah Kiat, for example, may snap up bargain-priced machinery or plants. But the natural environment is not necessarily a beneficiary. In 1993, the integrated forestry-paper firm International Paper had to balance its paper operations' losses by increasing timbercutting to a rate which the company itself admitted was 'unsustainable'. The Canadian giant Avenor, meanwhile, received state subsidies to open an idled mill (Penna 1994; *FT* 9.2.1995, 17.2.1995; *TN* 5.1994; Olsson 1995).

Exacerbating the industry's chronic tendency toward overinvestment and periodic overcapacity is the zeal of many national bureaucracies — and some of their consultant advisers — in promoting pulpwood or pulp as a foreign-exchange earner. Particularly in Southern countries, a pulp export industry is often advertised as being capable of dynamizing the economy and creating rural and industrial employment. In some countries, it is also promoted as a means of centralizing control over land, taking over small farmers' holdings, or even as an agent of 'reforestation'. In Thailand, government targets for pulpwood plantation acreage regularly overshoot even the most optimistic market demand projections by 1,000 per cent, and the ambition of some Indonesian government officials to turn the country into the 'world's biggest pulp producer' have contributed to an even more overheated atmosphere there. One Western machinery supplier has claimed that many prospective Indonesian investors 'don't know what they plan to do with their pulp', joining experts who worried in the early 1990s that the regional Asian market was 'undeveloped', uncertain, and oversupplied (Sargent 1990a; *Paper* 4.2.1992; Dench 1993; ADB 1993; WALHI and YLBHI 1992).

Concentration and liberalization

The cycles endemic to the pulp and paper industry tend to lead to renewed efforts to increase capital intensity and concentrate production in fewer hands. When, during slumps, Northern firms lose market share of basic paper grades to Southern producers, they are forced to seek out and seize miniscule competitive advantages which might enable them to get rid of their overproduction. Jobs are shed, still bigger machines sought, and attempts made to diversify into new, higher value-added products or concentrate on fewer niches.

As corporations struggle to cut costs, hold market share, prevent overinvestment, find and control markets, create demand, and cope with globalization, they are often pushed into new mergers and cartels. Rapid upturns, when they come, provide acquisition-hungry companies with flushes of cash they can use to buy out vulnerable rivals. Economy measures meanwhile induce many companies to link themselves more tightly to timber producers, since large-capacity, wood-based mills rapidly create shortages of timber unless combined with sawmill operations or plantations (JPA 1994, *PPI* 7.1994, 7.1995, Rajesh 1995). Links are also increasingly sought with paper-product manufacturers and paper customers.

Thus in the last decade the top 30 paper companies have increased their proportion of world production dramatically (Higham 1995). In 1995 alone, the US's Kimberly-Clark took control of Scott Paper, Sweden's SCA bought Germany's PWA, the US's International Paper gained control of New Zealand's Carter Holt Harvey and initiated moves to acquire Federal Paper Board, Canada's Canfor launched a bid for Slocan, and Jefferson Smurfit took over Saint-Gobain Paper-Wood. In Finland, in addition, mergers were announced both between Enso-Gutzeit and Veitsuluoto (to create ENSO) and between United Paper Mills and Kymmene (to create UPM-Kymmene); the latter marriage will result in Europe's largest pulp and paper company. In Japan, meanwhile, the share of national pulp production of the top five firms rose from 40 per cent in 1960 to 60 per cent in 1990, and in a further phase of what Alastair Graham calls 'mutual cannibalism' in 1993, Jujo combined with Sanyo Kokusaku Pulp to create Nippon Paper, while Oji merged with Kanzaki, Honshu and Chuetsu to form New Oji. In 1994, the two new firms took over the No. 2 and No. 3 positions in world pulp and paper sales (Whitham 1994; *PPI* 7.1995).

Pira International, an industry research centre, predicts that the share of world paper and board tonnage held by the largest companies could be well over 80 per cent by 2005, up from 65 per cent today (*FT* 13.7.95).

In addition to helping companies concentrate production in the most 'efficient' mills, mergers are likely to improve corporate ability to streamline organizations, sell non-strategic assets, control prices and wages and generally minimize risk at the expense of smaller economic actors. Local communities, however, may find that protecting their land, water and air from the actions of one of the new conglomerates is even more difficult than holding the line against their predecessors.

Industry concentration both feeds on and creates pressures for centralized international structures of economic control such as the European Union, the North American Free Trade Agreement (NAFTA), and the General Agreement on Tariffs and Trade (GATT). Large-scale producers expect NAFTA and GATT to open trade routes and foreign investment opportunities for them, while the European Union is helping the biggest pulp and paper firms there to gain access to wood, markets and production sites across the region, transforming it into a single economic unit with greater global power. Tariff reductions associated with a global market, meanwhile, tend to force all manufacturers into competition with their counterparts in the most cheaply-producing countries, causing the consequences of over-investment in one region to reverberate worldwide.

Economic liberalization is also helping capital intensive, wood-based manufacture to spread into previous strongholds of small-scale, agriculture- or farm forestry-based industry such as India and China. In India, for example, as lower import duties on paper begin to threaten the highly-protected, flexible, decentralized and labour-intensive local paper industry, Indonesia's Sinar Mas group has announced plans to set up a new mill twice as large as any other in the country, fed by surplus hardwood kraft pulp from its operations in Sumatra. The hope is that the mill's output will help create an Indian demand for a 'high-quality' wood-based sheet which will spur the development of other such mills, which of course will in turn require cheap new timber sources in natural forests or plantations (Bayliss 1995; cf. *DTE* [Delhi] 31.8.1995).

Economic risks in South and North

The highly cyclic nature of the pulp and paper industry creates risks for investors everywhere. But because the costs of installing world-class pulp or paper machinery are as high in the South as in the North, Southern countries who are pushed into investing in the industry risk a higher ratio of their resources than do their Northern counterparts.

These hazards are exacerbated by additional uncertainties. Due to the scale of today's world-class mills, unanticipated increases in pulp capacity due to the whims of single manufacturers in Brazil, Russia, Indonesia or elsewhere have great potential for upsetting the market. In addition, while the per capita paper consumption of some of Asia's newly-industrializing countries has exploded in recent years, it may stagnate if an economic recession hits. And while large countries with low per capita consumption such as Indonesia and China may increase world demand greatly if incomes rise, the timing of such changes is uncertain. It is conceivable, too, that if enough Northern producers are driven out of business by new Southern competitors, the price umbrella they provide may disappear (Bazett 1993, *PPI* 7.1994).

Environmental activism adds another unpredictable factor to the equation. Forest and land conservation movements affect demand for and supplies and prices of pulpwood, while recycling movements and legislation can unexpectedly increase global supplies of waste paper. Environmental legislation regulating manufacturing processes also has a powerful effect on industry spending. The extent of the need to invest in machines capable of meeting consumer demand for chlorine-free paper is another difficult-to-assess factor, and one which is especially relevant to the prospects for Southern plantations, since chlorine-free papers require a high proportion of hardwood fibres (Bazett 1993, ADB 1993).

The increasing reliance on plantations which has followed on from depletion of old growth forests creates additional uncertainties for pulp and paper producers in both South and North. Even fast-growing plantations require especially long planning horizons of 10-15 years, due to the time it takes trees to mature, during which time any number of things may change. Large monoculture plantations are prone to pests, diseases and fire, and productivity may well dwindle after the first rotations (Good, Lawson and Stevens 1993). Moreover, it is notoriously difficult to estimate how much competition pulpwood plantations will have from natural forests in the Pacific region in the future. Depending on what probabilities are given for the future economic availability of pulpwood from remote conifer forests in Siberia and Yakutsk, and for the demand from the Chinese market, projections have varied from a surplus of 20 million tonnes a year to a deficit of 60 million tonnes. According to former FAO forest economist Alf Leslie, such 'inevitable uncertainties dwarf any market growth analyses based on stable market factors like population and income growth.' As the next chapter indicates, uncertainties about raw material availability have proved a particularly powerful influence on the Japanese industry, propelling it into ventures all around the Pacific basin (Graham 1994, *PPI* 1.1994).

Consumption and demand creation

Since the Industrial Revolution, owners and managers of Western pulp and paper industries have struggled to reorganize society in ways friendlier to themselves through large-scale mechanization. As this chapter has suggested, the style of mechanization they have followed is closely tied up with economic cyclicity, concentration of skills and of wealth, the spreading of risk, globalization, and deforestation. As will be explored further in later chapters, it is also associated with a pattern of heavy pollution, soil degradation, widespread dispossession, and massive subsidization by the public sector. Despite the crises growing out of this highly-centralized type of social organization, the pulp and paper industry as yet shows few signs of moving on to a more decentralized, smaller-scale system of specialized production.

One of the most important effects of the strategy of large-scale mechanization and mass production followed by today's mainstream pulp and paper industry is on how paper is used by ordinary people. Industrialists who have to sink huge sums into complex production technology have a strong incentive to try to predict and control prices. As economist John Kenneth Galbraith (1972) has long pointed out, that entails attempts at managing demand. Such attempts are all the more imperative in an industry subject to the influence of savage boom-and-bust cycles, rapid cultural and technological changes, and unpredictable environmental pressures on production and consumption. Thus top European paper executive David Clark (1994) recently told his colleagues,

> we shall have to fight for our future and create our own growth. . . . total demand has to be stimulated. The alternative, to do nothing, could produce a static or even declining demand with serious implications for the industry, its reputation, its technology and the quality of the people it attracts.

It is hardly surprising, therefore, that today's mainstream paper industry is constantly promoting new uses for paper in the overconsuming markets of the North as well as seeking new outlets for its products in growing markets such as that of East Asia. This dynamic inevitably sets a highly-centralized and -mechanized industry at odds with environmentalists and others concerned about overconsumption.

Stimulation of paper demand is, of course, nothing new, and is not something the industry has to undertake alone. Since the 1800s, new processes and commodities — ranging from lithography, paper shirt collars, slippers, cups, building materials, bags, toilet paper, drinks cartons, nappies, supermarket packaging, export packaging, fax paper and word-processing computer technology — have been embedding paper use ever more thoroughly into business and household activities. But the scale and intensity of demand creation has recently reached new heights. To take one example, TetraPak, partly financed by the World Bank, recently opened a disposable carton factory in Hungary that caused the country's deposit bottle system to collapse in a matter of weeks (Fairlie 1992). New technology such as fax machines and laser printers, as well as the growing roles of service and administration in Northern economies and the burgeoning supermarket system, with its vast needs for packaging, also stimulate paper consumption.

While industry sometimes claims when speaking with uninformed audiences that the driving forces behind increasing paper demand are growing literacy, surging population, and the need for schoolbooks and education, the figures tell a different story. Over 40 per cent of paper production in 1991 was used for packaging and wrapping, 13 per cent for newsprint, and less than 30 per cent for printing and writing (IIED 1995, cf. Bazett 1993). In the US, according to Worldwatch (1994), it is the 'growth of advertising and the automation of office equipment' — including, paradoxically, computers, which were supposed to have introduced the 'paperless office' — which have driven growth in paper consumption:

> Sixty per cent of space in American magazines and newspapers is reserved for ads, while some 52 billion assorted advertising pieces — including 14 billion mail order catalogues that often go straight into the trash — clog up the US post office every year.

Explosive increases in cardboard and paper packaging in the last few decades have also been largely fuelled by the desire to advertise the products within, and even increases in demand for newsprint are primarily due to advertising (Bazett 1993). Overall, paper consumption is growing fastest in coated lignin-free and coated mechanical varieties, which are used for glossy magazines and colour advertising. Yet per capita newsprint use also continues to surge — from 24.5 to 32 kilogrammes between 1980 and 1991 in Britain alone *(G* 25.11.1995). In Southern countries, meanwhile, increases in paper consumption are closely tied to demand for packaging by industry (often export industry), advertising and the computerization of businesses, and the rise of consumerism. In Thailand, for example, one of the world's fastest-growing export economies, packaging absorbs more than two-thirds of production. As trade becomes more and more globalized, it is likely to entrench packaging demand even further (Kroesa 1990, Soulas 1994, *P&PA* 11.1994).

TABLE 2.4
Paper consumption, selected countries, 1994

Country	kg per capita	Country	kg per capita
US	332	Brazil	28
Hong Kong	233*	Bulgaria	23
Sweden	232	Turkey	22
Japan	231	Uruguay	22
Taiwan	224	Russia	20
Canada	221	China	20
Singapore	218	Philippines	12
Finland	217	Egypt	10
Germany	201	Indonesia	10
UK	197	Lithuania	9
New Zealand	184	Saudi Arabia	9*
Australia	167	Peru	8
Italy	143	Dom. Rep.	8
South Korea	137	Kenya	6
Spain	125	Nicaragua	4*
Slovenia	100	India	4
Ireland	98	Bolivia	3
Malaysia	82	Nigeria	3*
Portugal	74	PNG	2*
Costa Rica	55	Bangladesh	2
Argentina	45	Viet Nam	1
Chile	42*	Nepal	1
Mexico	42	Burma	1
Latvia	40	Tanzania	<1*
Venezuela	39	Congo	<1*
Poland	38	Ghana	<1*
South Africa	38	Lao PDR	<1*
Thailand	35	Mali	<1*
		World	48 *

*Apparent per capita consumption.

Source: PPI 7.1995

Tying demand for paper to a broad range of economic activities outside educational publishing is one of the factors which has allowed world per capita paper consumption to expand indefinitely. Rising from

.01 kilogrammes yearly in 1910 to 15 kilogrammes in 1950 and nearly 48 kilogrammes in 1994, it shows no signs of levelling off (Worldwatch 1994, *PPI* 7.1995). The industry's 'efficiency' can no longer be described plausibly as, say, 'efficiency in producing the medium for the books and newspapers which society needs', but is increasingly merely an abstract ability to produce as much paper as possible as cheaply as possible. Unsurprisingly, per capita paper consumption has very little relation to literacy levels (see Table 2.4).

According to industry figures, in 1993 the South and Eastern Europe, with nearly 84 per cent of the world's people, consumed less than a quarter of the paper and board produced worldwide, while the North, plus the Asian 'tigers', with just over 16 per cent of the world's population, accounted for over three-quarters. Worldwide, wood fibre consumption is skewed even more than consumption of other commodities. US citizens, for example, consume 386 times as much pulpwood per capita as Indians, while 'only' 43 times as much oil (G 6.9.1994). Almost half of the world's paper production is consumed by 460 million people in the US, Japan and Germany, with the US alone consuming more than Japan, China, Germany and the UK (the next four top consumers) combined (see Table 2.5) (TRN 1993; IIED 1995; Ozinga 1994; Soltani and Whitney 1995; Dudley et al. 1995).

TABLE 2.5
World's top consumers of paper, 1994

Country	Apparent consumption (million tonnes)	Per cent of world consumption
US	85.8	32
Japan	28.8	11
China	24.3	9
Germany	16.3	6
UK	11.1	4
France	9.7	4
Italy	8.3	3
Canada	6.2	2
S. Korea	6.0	2
Spain	5.0	2
Taiwan	4.7	2
CIS	4.6	2
Brazil	4.6	2

Source: PPI 7.1995

Significantly, the single-year *increase* in per capita consumption in Sweden and the US between 1993 and 1994 was five times the *total* current per capita consumption of India and double that of Indonesia (*PPI* 7.1995). While the most rapid recent growth in consumption has been in South Korea, China, Indonesia, Malaysia, Thailand, Taiwan, Singapore and Chile, it is the North which will continue to dominate world consumption figures. The US Forest Service suggests, in what unfortunately may well be a conservative estimate, that annual per capita consumption in the US will reach 472 kilogrammes by 2040 (Smith forthcoming).

Overall, paper consumption has climbed from less than 15 million tonnes in 1910 to over 268 million tonnes in 1994, with the paper consumption of Japan alone doubling in two decades. According to industry figures, consumption growth from 1988 to 1996 will equal total consumption globally 30 years ago (Ionides 1994). Although rates of paper consumption can change, and have changed, remarkably quickly, the UN's Food and Agriculture Organization strives to project such trends as global destiny, predicting that paper consumption will increase by 80 per cent by the year 2010, with more than 60 per cent of this increase to take place in Europe, North America and Japan. Roger Olsson of Taiga Rescue Network calculates that even if recycling rates increase from the current 35 per cent or so to 50 per cent by 2010, world pulpwood consumption will still increase by 33-41 per cent over 1990 levels (Olsson 1995, *BP* 22.2.1995).

Interestingly, while per capita demand for paper continues to rise, per capita demand for sawnwood is lagging, with less construction timber and fewer pitprops needed. Between 1955 and 1985 the volume of sawnwood consumed per unit of Gross National Product halved, while paper consumption followed GNP closely. Accordingly, the composition of demand for industrial wood has changed radically during the 20th century. Between the 1940s and the 1980s, the ratio of pulpwood to saw-logs increased from about 1:4 to between 1:2 and 1:3. Although FAO and the World Bank anticipate annual increases in paper demand to 2000 on the order of 2.7–2.9 per cent, they expect sawnwood demand to increase only 1.0–1.5 per cent. According to Michael Bazett, one industry analyst, growth in production of sawlogs is now close to zero. Increases in wood consumption, therefore, are due more and more to paper consumption (Mather 1990, Gauthier 1991, Bazett 1993).

Chapter 3
Emergence of a Global System

A global market

In the 1950s, most production of fibre, pulp and paper was for domestic markets, and few large producers and consumers of paper imported raw materials from other continents. Although there was considerable trade of newsprint and some other grades across the borders of Southern countries, the bulk of international wood fibre and pulp trade was between Canada and the US and among European countries, with Finland, the USSR and Canada being the leading raw fibre exporters.

Today, the industry as a whole still produces mainly for domestic markets, and the North continues to dominate pulpwood production (see Table 3.1). Only nine per cent of fibre production, 17 per cent of pulp production, 23 per cent of paper production, and 16 per cent of waste paper production crosses national borders. The overall export market for wood pulp in 1993 was valued at little more than US$11 billion and for paper at not much over $43 billion.

Nevertheless, international trade in wood fibre has quadrupled since 1960, while the world market for pulp has grown fivefold over the past 40 years. Trade in paper, including waste, has also grown enormously. As Russian logs travel to Finland, chips and pulp from Canadian forests are shipped across the Pacific to feed Japanese paper mills, Indonesian pulp and paper surfaces in the Middle East and East Asia, and vast amounts of the US's discarded paper are exported to Mexico and the Far East. As industry leaders point out, the pulp and paper economy has become global in just a few decades, with plantation, pulp and paper businesses nearly everywhere in increasingly active competition with firms from distant countries (Dudley 1992, Hagler 1993, IIED 1995).

Thus the wood fibres in a sheet of paper in Japan or Europe today may well originate from trees on five or six continents. To take perhaps the most extreme example, some 74 per cent of the wood fibres found in Japanese paper and board come from trees grown abroad. Between 45 and 50 per cent come from the US and Canada, over ten per cent from Australia and New Zealand, around seven per cent from Chile, over three per cent from Brazil, perhaps one to two per cent each from Finland, South Africa and Indonesia, and smaller amounts from

Russia, China, Thailand, Fiji, Sweden, Norway and other countries. The average wood fibre found in a sheet of Japanese paper or board, in other words, has travelled more than 6,000 kilometres from its point of origin (estimated from JPA 1994, *PPI* 7.94, UN 1994).

TABLE 3.1
World's top raw fibre producers, 1991

Country	Per cent of world production
US	34
Canada	11
CIS★	10
Brazil	7
Sweden	6
Finland	4
Germany	3
Japan	2
China	2
France	2
Spain	2
Australia	2
Portugal	2

★Former USSR minus the Baltic states

Source: FAO

In aggregate terms, most fibre, pulp and wood trade is still North–North. In 1991, nearly 83 per cent of wood fibre trade and 77 per cent of pulp trade was among the long-industrialized countries, with exports from South to North accounting for only 13 per cent of the fibre trade and seven per cent of the pulp trade. The US and Canada still produce close to half of the world's pulpwood and pulp and account for more than 25 per cent of world fibre exports, 55 per cent of pulp exports, 31 per cent of paper exports, and over 44 per cent of waste paper exports. Many of the biggest paper corporations, in addition, are US firms also producing timber, sawnwood and plywood (IIED 1995, *PPI* 7.94).

Yet this numerical Northern dominance conceals the deep impact the globalization of the pulp and paper economy has had on the South. As fibre, pulp and paper are shipped over greater distances, often by sea, the South has become more integrated into the global market than ever before, with several Southern countries moving up to the status of world-class exporters and importers within the last decade. (See Tables 3.2, 3.3, and 3.4.) The still small numerical ratio of

Southern to Northern fibre in the world market, moreover, says nothing about the rise in absolute amounts traded, the extent to which new pulp capacity in Southern countries is aimed at export (see Table 3.5), or the large changes the globalization of the market has brought about in specific Southern rural societies.

TABLE 3.2
Top exporters of pulpwood and chips, 1991

Country	Per cent of world exports
US	21
CIS	15
Germany	10
Australia	10
Chile	10
France	7
Canada	4–5
Hungary	2
Norway	2

Sources: Estimated from IIED 1995; United Nations import–export figures; USDA 1994.

TABLE 3.3
Top exporters of pulp, 1994

Country	Estimated per cent of world exports
Canada	30
US	17
Sweden	10
Brazil	7
Chile	5
Finland	5
Portugal	4
Russia	3
New Zealand	2
Spain	2
South Africa	2
Norway	2

Sources: Estimated from Stefan 1995; IIED 1995; PPI 7.1994, 7.1995; UN import–export figures; Wright 1993. Indonesia will soon be a prominent exporter.

TABLE 3.4
Top exporters of paper, 1993

Country	Per cent of world exports
Canada	34
Finland	26
Sweden	21
Austria	6
Norway	4
Brazil	3
Indonesia	1

Sources: PPI 7.1994; UN *import–export figures.*

In 1992, for example, Brazil, the South's preeminent pulp and paper exporter, sent around three-quarters of its approximately US$750 million worth of pulp exports to Europe and North America — the US and Belgium taking the lion's share — with another quarter going to Japan and the rest of East Asia. Only around two per cent went to Latin America. Brazil's nearly $600 million in paper and board exports were spread over even more countries, with Argentina, Italy, the UK and Nigeria leading the pack. Some 35 per cent went to Europe, 30 per cent to Latin America, 14 per cent to Africa and 12 per cent to the Middle East. At the same time, the country imported over $210 million worth of paper, mainly from Canada, Finland, and the US (UN 1993).

Chile, meanwhile, sold 96 per cent of its reported 1993 wood chip exports of $137 million to Japan, while the bulk of its $150 million in log exports, including pulpwood logs, went to South Korea (56 per cent), Turkey (19 per cent) and Japan (18 per cent). The country divided 80 per cent of its $468 million in pulp exports between Asia and Europe, with only 15 per cent going to Latin America. Leading buyers included Belgium/Luxembourg, taking 16 per cent, Japan with 12 per cent, and South Korea, Italy, Germany, China and the UK with around seven per cent each. While Chile did export most of its paper and board to its own region of Latin America, with a quarter of its $77 million in exports going to Argentina alone, the country also imported $143 million worth of paper and board, mainly from the US, Brazil and the Nordic countries (USDA 1994, UN 1994). (See Table 3.4.)

The continuing statistical dominance of North–North trade, moreover, can obscure the way in which parts of the North — including parts of Northern Alberta and the southeastern US — are being transformed by the globalization of the industry into new 'Souths within the North'. In Iberia, for example, state subsidies have helped ensure that woodlands useful to local economies are replaced with pulpwood plantations for the international market. In 1955, pulpwood constituted only seven per cent of commercial wood in Spain, but by the late 1980s, over 50 per cent. Pulpwood trees now cover five million hectares in Spain, other hardwoods only 1.9 million. Erosion and pest infestations have increased, while biodiversity has been cut, commons enclosed, and rural areas depopulated. Spain's 1985 entry into the European Community only reinforced this process. Since EC rules 'assign' milk production to Northern Europe, many former milk-producing areas in Spain have switched to pulpwood plantations. European Structural Funds aimed at integrating Spain's regions into the larger European economy have subsidized the mechanized clearing of *maquis* vegetation for 'reafforestation' — a process which hastens erosion and decimates wildlife (Coordinadora Extremena de Protección Ambiental 1992).

TABLE 3.5
Ratio of exports to production of chips, pulp and paper, selected countries (per cent)

	Chips (1993)	Pulp (1994)	Paper (1993)
Australia	100	0	13**
Brazil	0	34	27
Canada	5	36	66
Chile	52	71	29
China	59	0	1
Finland	6	15	92
Indonesia	33*	***	26
New Zealand	14	48	31
Portugal	7*	68	42
South Africa	8*	28	14
Sweden	3	26	80
Thailand	na	20	4
US	8*	9	6

*Estimated from 1991 FAO figures for pulpwood and particles.
**Based on 1992 UN export–import figures.
***Indonesia plans to export over 60 per cent of its pulp production by 1997.
Note: Exporter status does not imply self-sufficiency in all grades. For example, although Chile and the UK export some grades of paper they are net importers of paper, partly because they import grades which they do not produce.

Sources: Estimated from Stefan 1995; PPI 7.1994, 7.1995;
USDA 1994; UN trade figures; Wright 1994; ABECEL n.d.

Globalization of production has been accompanied by an increasing regionalization and globalization of investment and ownership. Today Swedish firms operate throughout Europe; New Zealand-based Fletcher Challenge holds firms in North America; Australia's Amcor manages operations in China, Europe, Japan and Southeast Asia; Japan's Daishowa is in the forests of western Canada; Finland's partly state-owned Enso is investing in Kalimantan; the US's Weyerhaeuser is moving into eastern Siberia; German and Finnish firms are in Galicia; British companies are in Brazil and Swaziland; and Japanese firms are planning moves into China. While India's Ballarpur operates in Thailand, Thailand's own Siam Pulp and Paper is investing in mills in the Philippines, and Indonesia's Indah Kiat and Raja Garuda Mas are putting money into mills in Bombay and Sarawak.

Export destinations

Where is all the exported wood fibre, pulp and paper going? The answer depends on many factors, including how close the exporting country is to an importer. Some of the most voracious importers for which statistics are easily available are listed in Table 3.6.

TABLE 3.6
Fibre, pulp and paper imports, selected countries, 1993

COUNTRY	IMPORTS (million metric tonnes)			
	rough wood*	chips and particles	pulp	paper
Japan	16.8	11.3	3.3	1.3
US	-	0.9	4.9	na
Finland	4.2	0.4	-	0.3
Sweden	3.6	0.4	0.2	0.3
Italy	3.6	0.3	2.6	3.0
South Korea	7.3	0.5	1.5	0.5
Germany	0.9	0.1	3.6	6.8
France	1.0	0.4	2.0	4.0
Netherlands	0.4	0.2	0.7	2.5
China	2.5	0.1	0.5	0.2
Hong Kong	0.4	-	-	2.4
UK	na	na	1.6	6.0

*Includes non-pulpwood.

Sources: Estimated from United Nations export–import statistics, IIED 1995, Rice 1995, Hagler 1995.

One of the ironies of the globalization of the pulp and paper industry is that over half of all world exports of raw fibre for paper-making are shipped to three of the most heavily-forested countries in the world — Japan, Finland, and Sweden — all of which make it a matter of national pride to maintain plentiful domestic tree cover. In 1991 these three nations absorbed 40, nine, and eight per cent of the trade respectively; Belgium/Luxembourg ranked next, with seven per cent (Hagler 1993, IIED 1995).

Importing the most pulp, meanwhile, are the US, Germany and Japan, with 17, 15 and 11 per cent of all imports, although these three countries are also among the biggest world producers. The top paper importers are the US, with 19 per cent of world trade, Germany, with 13 per cent, and the UK, with ten per cent. Taiwan and South Korea, meanwhile, import the most waste paper, with ten per cent of world trade each, followed by The Netherlands with eight and Mexico and Canada with seven each.

The biggest South–North flow of raw fibre is that of wood chips from Chile to Japan, and the largest South–North flow of pulp from Brazil to the US. The biggest North–South pulp trade, meanwhile, is from the US to South Korea, China and Mexico and the biggest North–South paper trade from the US to China (IIED 1995).

Overall, the global fibre, pulp and paper economy can be divided roughly into Atlantic and Pacific trading regions. In the Atlantic region, which is the focus of world trade in paper, pulp and paper production is concentrated in North America and the Nordic countries, with growing links to wood or pulp sources in countries such as Brazil, South Africa and Congo. In the Pacific region, most industry trade consists of logs, chips or pulp sold to Japan and, to a lesser extent, China and the Asian 'tigers'. While in the Atlantic region, there is a growing shortage of softwood for pulp, the Pacific region has seen a shortage of hardwood for a decade, although with new eucalyptus plantations this may ease (Hagler 1993).

Causes of globalization

The globalization of pulpwood, pulp and paper production is due to a combination of many factors. First, the industry's culture and institutions, like those of industrial capitalism generally, tend to be oriented toward unlimited accumulation, economic expansion, and consumption growth. When resources in one area can no longer be tapped to feed this growth due to economic, ecological, geophysical or political reasons, but are accessible elsewhere, pressures for globalization increase. Growth in paper consumption in Japan after

1960, for example, has been largely dependent on, first, finding West Coast North American alternatives to expensive or inaccessible local fibre sources, and, second, as North American sources become less economically, politically and biologically accessible, on finding further alternatives in other regions of the world (see below). In a world of skewed power relations and resource distribution, globalization thus offers a way of preventing local circumstances — whether they are classified as economic, biological or political — from limiting local consumption. Just as economic growth externalizes costs to the future through debt and discounting, the global economic expansion which accompanies this growth externalizes costs to 'outlying' regions.

Industry- and country-specific factors have helped lay out the particular paths toward globalization which wood fibre, pulp and paper have taken. Among these are the cultural and political elements mentioned in chapter 2, which, over time, have locked the industry into reliance on big machines, reinforced the need for a huge forest resource, and contributed to recurring rhythms of over-investment, recession, concentration, and demand stimulation. It is partly these rhythms which define the quest for more trees and paper buyers which has pushed the industry to expand outside its traditional centres.

Other forces behind globalization include the North American industry's long-standing practice of depleting old-growth forests and investing in new pulping technologies and biotechnology research instead of replanting extensively. As native coniferous forests are exhausted, environmental activism rises, further cutting into pulpwood harvests. Union Camp Corporation estimates that, partly due to environmental concerns, sale of timber from public lands in the Pacific Northwest will be 64 per cent lower in the second half of the 1990s than it was a decade earlier, while wetlands legislation could reduce acreage available in the southeastern US by ten per cent. That amounts to the equivalent of the raw materials needed for 30 world-class 400,000 tonne-per-year pulp mills. Pressures to conserve old-growth forests in the Nordic countries, Australia and Chile have also been significant. The Finnish industry, meanwhile, was jolted in the early 1990s by the disruption of Russian exports of wood, especially hardwood, due to the breakup of the Soviet Union (Marchak 1992, *Know-How Wire* 1993: 5, *JP&P* 30 (1), Ozinga 1994, McClelland 1994, Hagler 1993).

Such developments are leading the industry not only to look toward as-yet unexploited old-growth forests, but also to plan for increasing reliance on plantation fibre. Plantations are especially attractive in that they promise to be able to furnish exceptionally uniform raw material more quickly than natural forests and on a smaller land base, avoiding conflict with other land uses. While industrial plantations currently account for only about 15–30 per cent of world demand for pulpwood (trade consultant Robert Hagler estimates that only 11 per cent of 1993 global pulp production was based on fibre from plantations of exotic trees), these ratios are bound to rise, given deforestation, decreased availability of extensively-managed forests, the limitations of recycled fibre, and the resistance of much of the industry to non-wood raw materials. Already by 1990, 95 per cent of Chile's industrial wood production, 93 per cent of New Zealand's, and 60 per cent of Brazil's came from plantations. By the year 2030, Indonesia plans to increase the share its plantations take in production of industrial wood from the present 20 per cent to 80 per cent (Hagler 1994, 1995; Bazett 1993; IIED 1995; R. Wilson 1991, 1995; Stefan 1995; Pandey 1992).

The more industry shifts from natural forest to plantation pulpwood, the more incentive it has to move raw fibre production to the South. For one thing, although growth rates vary extremely widely from place to place and depending on the methods used, and are always higher in experimental plots than in large-scale plantations, fast-growing trees such as eucalyptus generally grow much faster in the South than commercial species do in the North, meaning both that they are available earlier and that less land is required for plantations. While the average growth rate of managed forest and plantations in the US is about 2.6 cubic metres per hectare per year, pine plantations in the South have been shown to yield from 5.7 cubic metres at certain locations in Madagascar to 30 in some Chilean plantations. Southern eucalyptus plantations have meanwhile yielded from around 1.5 cubic metres annually per hectare in Burundi to six in various locations in India all the way up to 70 under certain exceptional regimes in Brazil. Forestry consulting firm Jaakko Pöyry estimates that while 1.6 million hectares of replanted forest area in British Columbia's interior are required to feed a 500,000 tonne-per-year pulp mill, and 800,000 hectares in the Nordic countries, only 50,000 hectares are required in Brazil under ideal conditions. Bazett calculates that to feed a world-class mill, 1.3 million hectares of non-managed coniferous forest would be required, compared with 650,000 hectares of intensively-managed natural coniferous forest and 80,000 hectares of fast-growth plantations (Evans 1992, Pandey 1992, Axberg and Ståhl 1989, Bazett 1993).

Land is also cheaper in the South, particularly in big contiguous parcels. In many countries, for example, the state rents out nominally 'forested' reserves to plantation companies at well below market rates. In Indonesia, state land can be rented by plantation firms for about US$0.30 per hectare per year, in Thailand for around $2.50, and in Lao PDR for $3.00 (WALHI 1990, Bannan 1995). While land rental dominates the costs of tree planting programmes in the US, one company which plans to set up a plantation in Lao PDR plans to spend a mere six per cent of its annual outlay on lease of land (Moulton and Richards 1990, Bannen 1995).

All this makes for low wood costs. According to the Canadian consulting firm H. A. Simons, the cost in 1988 of producing one bone-dry metric tonne of hardwood fibre was only slightly over US$28 in Brazil, Chile and Argentina, $40 in the southeastern US, $49 in the interior of British Columbia, $102 in the Nordic countries, and approximately $154 in Japan. Softwood fibre production costs, according to the same company, were less than $28 in Chile but over $42 in the southeastern US, over $70 in Australia, and over $140 in Japan. In 1993, although competition with the South had forced Northern wood prices down, Brazilian and Indonesian hardwood fibre costs, and Chilean and Brazilian conifer costs, were still less than half those prevailing in the Nordic countries or the US's West Coast, and also less than in eastern and western Canada, Iberia, and the southern US. More recently, the gap has widened again, as North American and Nordic raw material costs have increased sharply (Graham 1994; *Know-How Wire* 1.93; Bazett 1993: 77, 92–3; Hagler 1994, 1995).

Such cost differentials are critical, since wood represents 40 to 70 per cent of the variable cost of making pulp, which is in turn the most important cost in making paper. As Robert A. Wilson remarks, 'Wood is *the* strategic driver in the industry . . . the key competitive differentiator'. It is thus often profitable to grow wood in the South even if the plantations are at a great distance from large paper markets. Barring political and economic turmoil, the move to the South for raw fibre is likely only to accelerate in coming years, as wood chips from the northwestern and southeastern US and western Canada become scarcer or more expensive (although Siberian softwood exports may also increase). Consultant Robert Hagler estimates that between 1990 and 2010, while the annual allowable cut of conifers in the US Pacific Northwest will decline from 100 to 70 million cubic metres per year, the yearly yield of eucalyptus plantations — situated mainly in the South — will surge from 82 to 132 million cubic metres and that of conifer stands in Australia, New Zealand and Chile will increase from 36 to 62 million cubic metres. Bruce Arnold, another US forest industry consultant, claims that industrial forest plantations could account for 50 per cent of the world's industrial wood production by the year 2000 (Wilson 1991; *Know-How Wire* 1.1989; Bingham 1995; *PPI* 8.1993, 1.1994; Hagler 1993, 1995; Wright 1993; *PP* 1.1995; Stefan 1995; Pandey 1992).

Discounted land is not the only subsidy encouraging the expansion of pulpwood plantations in the South. As later chapters will document, governments also help provide tax exemptions, low-interest-rate loans, low-cost labour and political repression. Hourly wages in Brazil, for example, are 20 per cent of those in Germany. Suppression of labour unions is meanwhile provided free of charge by the governments of many countries witnessing a plantation boom. As chapter 5 will argue, such subsidies are topped up through support by international agencies and even NGOs for infrastructure and for research and development programmes which disproportionately benefit the industry.

The move to plantations, and in particular plantations in the South, coincides with the growing acceptability to manufacturers of plantation fibre, especially eucalyptus. (See chapter 2.) Conversely, the more inroads plantation fibre makes into the industry, the more manufacturers will be encouraged to treat raw material as a factor whose composition can be manipulated and homogenized. Whereas the industry has previously been largely dependent on diverse types of wood waste, and thus has had to rely largely on manufacturing processes to ensure uniform quality in paper, it is now increasingly capable of reducing variability in the raw material itself. Plantation output can be homogenized through choice of species, site, inputs, spacing, provenance, hybridization, cloning, macro- and micro-propagation, and DNA analysis. Genetic engineering is also getting under way. As Robert A. Wilson and O. Fernandez Carro of Arjo Wiggins Appleton note (1992), 'the process of linking genes to tree, pulp and paper characteristics is now beginning'. Wilson (1995) adds:

> Like the agricultural revolution from the wild wheat of Mesopotamia to modern, high-yielding, disease-resistant wheat, the forest industry is facing a new age of merging natural forest species with agricultural experience into modern fibre-cropping systems. Tree species . . . are following the same path and improvements as wheat, corn and potatoes.

Dependence on such uniformity is likely only to reinforce dependence on plantations (Fernandez Carro and Wilson 1992, R. Wilson 1995, P. Wilson 1995, Griffin 1995, L. Wilson 1994).

Shifting pulp production

Some of the same incentives which encourage the industry to shift pulpwood production to the South also encourage it to build pulp and paper mills there. Cheap land in big contiguous swathes, for instance, is an advantage not only for plantation interests but also for pulp manufacturers, since state-of-the-art pulp mills tend to be huge and are thus most economically sited in the centre of large, compact raw-material catchment areas. Low labour costs are of course also attractive to pulp firms, as is the eagerness of many Southern governments to 'apply stimuli' to the industry. Looser environmental regulations provide still another attraction. In 1990, the pulp and paper industry in North America had to devote 54 per cent of its total spending on new plant to environmental measures, and in Western Europe, 26 per cent, while elsewhere the figure was only 10 per cent. New air and water rules enacted by the Environmental Protection Agency are set to make pulp and paper production still more expensive within the borders of the US; International Paper claims that the rules will cost it alone more than $1 billion in capital improvements over a three-year period (Soulas 1994, McClelland 1994, Van Hook 1994, *FT* 9.2.95).

Small wonder, then, that Southern-produced pulp can be as cheap, relative to that of the North, as Southern-produced wood. In 1993, for example, bleached hardwood pulp cost only US$78 a tonne in Brazil but $156 in eastern Canada and $199 in Sweden. New regional and global trade agreements such as GATT are making it easier for the industry to take advantage of such cost differentials by shifting production to the South (Judt 1994, Hagler 1995). The consultant firm Hawkins Wright predicts that of all the major kraft market pulp capacity increases expected between 1994 and 1997, some 77 per cent, or over 3.6 million tonnes per year, will be sited in the Southern countries of Indonesia, Brazil, Thailand, South Korea, Chile and Morocco. Over 98 per cent of the hardwood kraft market pulp capacity increases expected over this period will be in the South (Bingham 1995).

Pulp, moreover, has more value added, and, when dried, is more efficient to ship than logs or wood chips, which are up to half water. One dry tonne of hardwood pulp is roughly equivalent to 2.5 tonnes of hardwood chips; while it costs US$150 to ship, from Chile to Japan, enough softwood to make a tonne of dry pulp, it costs only $55 to ship the pulp made from that softwood. Thus plantation sites and mills have been integrated in the export sectors of Brazil, Indonesia, and other Southern countries. After 2000, according to Robert Hagler, 'increased pulping capacity in wood-producing regions and more joint ownership of this capacity by producers which formerly imported raw wood' will cause international wood fibre trade to level off and eventually to decline (Bazett 1993; IIED 1995; Hagler 1993; Whitham 1994).

While the cost of capital for mills tends to be higher in the South than the North, Southern countries can offer compensations which are often more than adequate, including soft loans from multilateral development banks. In order to be able to reschedule the debt service on such loans, Southern producers are often forced to reduce prices to keep orders — and foreign exchange — coming in. That pushes paper prices down worldwide. Hard-pressed to compete and hold market share, some Northern firms may be driven to subcontract out some of their basic production to their Southern counterparts. Thus the label of a Northern paper firm may be affixed to boxes of photocopy paper flowing out of southern Brazil or central Sumatra, while the firm's own mills move into production of speciality papers. This pattern helps explain why the newest, biggest mills in countries such as Brazil produce so heavily for export, while the task of supplying the domestic market is typically left to older equipment (Graham 1994, Oinn 1994).

Contributing to pressures to establish mills in the South is Northern firms' need to export pulp and paper machinery. In the early 1990s, for example, a severe economic recession engulfed Finland, a country particularly vulnerable to pulp and paper industry cycles due to the high 30 per cent share its forest industries contribute to GNP. What with heavy indebtedness, cost-cutting and layoffs by the forest industry, firms such as Tampella, Valmet, Sunds Defibrator, and Ahlstrom began to strive especially hard to find more Southern outlets. Helped by Finland's Premixed Concessional Credit Scheme and the Finnish 'foreign aid' budget, Finnish machinery exports to Indonesia surged from nil to over US$100 million between 1990 and 1993, while those to Thailand increased nearly fivefold, to almost $110 million. Some 18 per cent of Finland's machinery exports now go to the two countries, up from a little over one per cent in 1990 (*PPI* 1.1994; Finland National Board of Customs 1990-3, Rasmusson 1994, Ulvila 1994).

Nordic-country consultancies in forestry, engineering and pulp and paper manufacture are also eager to find more contracts abroad. Some ten per cent of Finland's professional foresters were reportedly unemployed in 1994, and many are eagerly seeking corporate- or 'foreign-aid'-funded jobs in the South. According to Ulf Rasmusson of World Wide Fund for Nature Sweden, Nordic consultants in Indonesia not only play a major role in the development of industrial forest estates but have also been involved 'in most of the major mill developments on the island of Sumatra' and 'perhaps the majority of the large mill projects on Kalimantan, . . . which will be the focus of much of the pulp mill development ahead.' With their ability, through inside connections, to appropriate 'aid' funds for commercial purposes, Nordic forest industry consultants are also invading Laos, Cambodia and Viet Nam in large numbers, pressuring local bureaucracies with money and plans in their attempt to repeat past industry 'successes' in Brazil and other countries. With some analysts predicting a stagnation in Nordic harvest volumes if large areas are planted in the South, pressures to export Nordic consultants may well increase further in the future (Rasmusson 1994, Bazett 1993).

One of the most important incentives for investment in new plantations and mills in the South, and particularly in operations which integrate the two, is the expected world-leading growth rate in Pacific Rim paper and pulp consumption over the next decade. By 1998, market pulp demand in Asia is likely to have outstripped that in North America. Some 45 per cent of the region's consumption growth over the next decades, according to FAO, will be attributable to Japan, which already accounts for half its paper use, although growth in the Asian 'tigers' and China will also be important. According to Arjo Wiggins Appleton, annual consumption of printing and writing papers in East and Southeast Asia will nearly double between 1993 and 2000, from 8.5 million to 16.5 million tonnes. Latin American paper consumption is also projected to grow, although less spectacularly, with demand for printing and writing grades expected to jump from three million tonnes in 1993 to 4.4 million tonnes in 2000. By comparison, Eastern European demand for printing and writing papers is projected to increase from 1.9 to 2.5 million tonnes (Wright 1993, Olsson 1995, Soulas 1994, Ionides 1994, Graham 1994, WALHI and YLHBI 1992, Soetikno 1993, Aurell and Jaakko Pöyry 1988, Pesonen 1995).

Investment in plantations, pulp and paper is moving not only from North to South and North to North, but also from South to South and even from South to North. The capital-surplus economies of South Korea and Taiwan, for example, are increasing investment in chip- and pulp-producing capacity not only in Southeast Asia but also in, for instance, the southeastern region of the US, partly to ensure secure supplies for domestic paper manufacturers. At the same time, a wood-hungry Chinese concern is looking to help build a pulp mill which would feed off native forests in southern Tasmania, and Malaysian and Chinese firms have joined Japanese and US counterparts in investing in recently-privatized state forest land in New Zealand (Graham 1994; Dudley, Stolton and Jeanrenaud 1995). South Africa's leading pulp and paper firm Sappi, meanwhile, controls companies in the UK, Germany and the US, while Mondi, the country's No. 2 producer, has interests in the UK and Portugal (*PPI* 3.1995).

Dynamics of globalization: the case of Japan

One striking demonstration of the regionalization and globalization of the pulp and paper industry is the expanding wood-fibre network centred on Japan. Japanese pulp and paper firms' exploitation of cheap foreign fibre sources has helped both to unharness national consumption from national supply and to contain the effects of environmentalist opposition at isolated points in the network.

Soon after the emergence of modern journalism and a modern wood-based paper industry in the latter part of the 1800s, Japan's paper companies, having depleted the limited native domestic conifer forests, were already eyeing overseas resources in Russia, China and elsewhere. After the Russo-Japanese War, Sakhalin was annexed and a large pulp production base set up there to take advantage of the peninsula's softwoods. Following victory in the Sino-Japanese war, Manchuria's forests were also exploited, as well as timber in Korea and Taiwan. By the time of the Second World War, Southeast Asia's forests were already being studied for pulp potential by Oji, and there was some exploitation of Southeast Asian mangroves as early as the 1950s. During the Cold War, the US government, seeing Japan as an ally against communism, offered Alaskan forests to a fibre-hungry Japanese consortium, causing a tremendous impact on the Tongass National Forest area and river system (Nectoux and Kuroda 1989, JATAN 1993, Kuroda 1995).

Between 1960 and 1970, Japan, although faced with a shortage of economically-available domestic raw materials, saw its apparent paper consumption surge from 47 to 121 kilogrammes per capita. Beginning in 1959 with the lifting of a quota system of foreign exchange for timber imports, the country began buying large amounts of more expensive foreign fibre, particularly sawmill wastes from western North American softwood forests. Still, shortages continued through the early 1960s, despite increased production in national and private forests. More and more was imported from North America, Australia, New Zealand, the Soviet Union, and — peaking in the mid-1970s — Malaysia. Between 1965 and 1975, the imported proportion of Japan's pulpwood supplies (and this does not include residues from sawmill processing of imported logs) jumped from three to 40 per cent. Between 1986 and 1993, overlapping with another period of rocketing consumption (Japanese paper consumption rates and oil prices appear to be inversely correlated), this ratio increased to nearly 60 per cent. In 1991 Japan was appropriating 40 per cent of world imports and 90 per cent of Pacific Rim imports (Marchak 1991; Ozinga 1994; Hagler 1993, 1995; JPA 1994; Penna 1992; Lamb 1992; Olsson 1995; IIED 1995).

Over the years, the form in which Japan imported its pulpwood changed from logs to wood chips — a more easily standardized international commodity. In 1964, the first of a fleet of high-sided, flat-hulled oceangoing vessels made especially for hauling the chips was constructed. Such ships now number nearly 110, and are 90 per cent Japanese-owned. Between 1955 and 1975, the proportion of Japan's pulpwood derived from chips rose from 0.2 to 74 per cent. Largely as a result of this shift, the proportion of total world wood fibre trade conducted in chips increased from ten per cent in 1960 to 54 per cent in 1990. Japan now accounts for 80 per cent of the world's wood chip trade, five times more than Europe and North America combined. As suggested above, the next stage is to import dried pulp rather than chips. Between 1991 and 1994 the share of pulp imports in Japan's apparent consumption rose from 20 to 26 per cent (Schreuder and Anderson 1988, JATAN 1994, Lamb 1992, Whitham 1995).

Since the 1960s, most of Japan's fibre imports have been western North American softwood chips. In about 1968, however, the country began to import plantation pine from New Zealand, rubberwood and mangrove wood from Malaysia, and, a couple of years later, Australian eucalyptus. In 1972, a pulpwood supply contract was signed with the Soviet Union, and tropical hardwood chips began coming out of a Honshu operation in Papua New Guinea in 1974. South Africa began shipping acacia and eucalyptus chips in 1976, and Indonesian mangrove forests began to be chipped for export to Japan about a year later.

Following a mid-1970s downturn in the Japanese industry, partially due to new pollution-control laws, a revalued yen, higher oil prices, and depressed industrial demand, three shocks led the country's industry to rush to diversify its overseas supplies even further. First, in 1979–80, interest rates increased in the US. This reduced housing starts, sawmill production, and thus mill residue surpluses available for export. US suppliers such as Weyerhaeuser, on which the Japanese industry had grown particularly dependent, imposed steep price hikes overnight. Prices did not remain high enough to make a shift to domestic supplies economic, and dropping oil prices in the mid-1980s encouraged the industry to turn to imports of softwood from countries such as Chile. With increased recycling, long fibres from these imported softwoods became particularly important to top up reused fibre mixes, but short-fibred *Eucalyptus deglupta* and *E. grandis* also began to be recognized as valuable for quality printing papers. By September 1981, Honshu, Oji, Jujo, and Kanzaki had teamed up in a joint firm aimed at securing new raw materials and unifying wood fibre research.

Second, in 1987–88, the Tasmanian government increased wood royalties, and a projected mill in the state threatened to absorb Australian eucalyptus supplies which had been going to Japan. These events led the Japanese industry to turn increasingly to Southeast Asia and the southeastern US. A third shock arrived in the early 1990s when logging was reduced in western North American forests, diminishing waste chip supplies drastically. This development was due to the forest industry's depletion of old-growth forests without sufficient replanting, accompanied by growing environmentalism affecting use of public lands. Chip exports from the western US declined almost 22 per cent between 1989 and 1992. The Japanese industry's worries about environmentalism in Australia and Chile have only reinforced its determination to secure varied raw material sources (*JP&P* 30 (1); *PPI* 1.1994).

By the late 1980s, amid a second big surge in domestic apparent consumption from 160 kilogrammes per capita in 1984 to 222 kilogrammes in 1989, Japan was importing wood chips at a steadily increasing rate not only from Australia, the western US, Canada, Chile, New Zealand, South Africa, the Soviet Union, Indonesia and Malaysia, but also from Fiji, Papua New Guinea, China, Thailand, Taiwan and the southeastern US. In addition, it was laying plans to secure further supplies from the interior of northern Canada, Vietnam, Argentina, Venezuela, West Papua, and other parts of Oceania. Siberia, which has always supplied some softwood to Japan, has also become a major target for further exploitation but must wait for infrastructure to be built, some of which may follow on from oil development. Japanese chip imports shot up from 7.1 to 11.8 million tonnes between 1987 and 1991, with total pulpwood imports nearly doubling between 1985 and 1991. During the period of overcapacity and falling profits in the paper industry in the early 1990s, Japan's Ministry for International Trade and Industry, rather than taking direct responsibility for reducing production and capacity, gave the industry yet further encouragement to invest in technical development, plantations and mills in Asia to take advantage of the region's resources, low costs and expected high demand growth (Whitham 1994, 1995). In 1993, 38 per cent of Japan's chips came from North America, 30 per cent from Australia and New Zealand, 15 per cent from Latin America, eight per cent from other Asian countries, and over one per cent from Fiji and Papua New Guinea. Over the next decade, hardwood supplies from Southeast Asian plantations are likely to replace a significant portion of imports from more distant regions such as Chile or the southeastern US. The Japanese paper industry's continuing foreign expansion has been critical in keeping its prices competitive with those of imports and in reducing its dependence on the nation's trading companies; in the future it may also be critical in the industry's attempts to exploit new markets (Penna 1992; Marchak 1992; Schreuder and Anderson 1988; P&PA 1993; Pöyry 1993; UN 1994; JATAN 1993; Hagler 1995).

As chapters 8–12 will show, one result of this drive to diversify sources was social strife and dislocation in Southeast Asia and Latin America. Another was changes in the ecological and social landscape in certain regions of the North. By 1993, for example, the southeastern US had become a major source of raw materials for Japanese paper, with Gulf of Mexico ports exporting over twice as many hardwood chips to Japan than their counterparts on the West Coast. As one Japanese industry figure commented, his nation had taken a 'new step to secure resources, that is, planting its own forests in other countries as long-term resource programs' (JP&P 30 (1), Hagler 1995).

Competition among such a large number of countries for the Japanese market helped keep prices low and provided a guarantee of steady supplies. By 1987, Australian hardwood was already cheaper in yen than domestic hardwood even when transport costs to Japan were added in, while US softwood was almost at parity with domestic softwood. The yen prices of these woods declined even further between 1990 and 1994, and Indonesian hardwoods and Chilean softwoods were still cheaper (Marchak 1992, Japan Paper Association 1994, *P&PA* November 1993, UN 1994, *JP&P* 30 (1), Hagler 1993, Whitham 1994).

Such low prices make it unlikely that the Japanese industry will turn from foreign to domestic supplies of pulpwood. While Japan is nearly two-thirds forested, its trees remain economically relatively unavailable for pulp. Instead of being managed by integrated corporations, forests are owned in small, fragmented parcels by families who band together in cooperatives and communities to manage them. Such owners are often not particularly interested in selling to the pulp and paper industry, which is located outside forested areas. Much Japanese forest land, moreover, is on steep slopes, is difficult to get access to, and suffers from poor management. While the government established ten million hectares of conifer plantations in response to heavy overcutting of hardwoods after the Second World War, it is extremely doubtful whether these trees, now over 40 years old and covering around a quarter of the nation's land surface, will be ready for harvest within the next ten years. Including such species as cedar and cypress, the plantations were in any case established mainly for lumber, not for pulpwood. Remnant hardwood stands, meanwhile, are being logged less and less, due largely to conservationist pressures (Marchak 1992, Cameron 1994, JATAN 1994, Bazett 1993, Kuroda 1995).

The increase in imports of wood fibre to Japan in the late 1980s and early 1990s was overwhelmingly in hardwoods. The postwar development of a hardwood pulp technology had enabled the exploitation of Japan's own broadleaved trees for paper, contributing to an increase in domestic hardwood harvesting in the 1960–72 period. But although domestic softwood harvests stayed roughly level through the 1970s and after, domestic hardwood supplies began to dwindle in the late 1960s, when beech forest felling, which had been increasing, began to trail off. Native hardwoods from countries such as Chile, Australia and Papua New Guinea, together with plantation eucalyptus and acacia from countries such as South Africa and Thailand, have satisfied much of the increased demand. As chapter 2 suggests, this shift has been enabled partly by eucalyptus' increasing attractiveness as a raw material for quality papers. By 1991, the proportion of hardwood to total chips had increased to 65 per cent, and by 1993 to

73 per cent (Graham 1994, JATAN 1993, Penna 1994, Marchak 1992, UN 1994, JPA 1994, Lamb 1992).

Historically, Japanese companies have tended to keep their pulp and bulk-grade paper mills close to home. Among other things, this enables the industry to take advantage of the economies of integrating pulp with paper plants. But what with pressures to secure cheap fibre supplies, to transport pulp rather than chips, and to accept foreign imports, Japanese overseas investments have increasingly included pulp and even basic-grade paper mills as well as plantations and chipping operations. In 1971, Oji and what was then Sanyo Kokusai Pulp helped establish the 700 tonne-per-day Carter Oji Kokusaku Pan Pacific pulp mill in New Zealand. Two years later, Nippon-Brazil Resource Development, a consortium of 20 Japanese paper companies and the Overseas Economic Cooperation Fund, joined together with Companhia Vale do Rio Doce in Minas Gerais, Brazil, to create Celulose Nipo-Brasileira (see chapter 7). The 40,000 tonne-per-year Tenma whiteboard mill was set up in Thailand shortly afterwards with the involvement of Sumitomo. By 1980, new pulp or paper capacity amounting to one million tonnes per year had also been built in Canada through ventures of Honshu (in association with the Mitsubishi general trading company), Daishowa (with Marubeni), and Oji (with Mitsui). Nippon Paper and Alaska Pulp Corporation (in association with Mitsubishi, Marubeni and Itochu) meanwhile helped build 800,000 tonnes of dissolving pulp and newsprint capacity in the US.

This trend accelerated in the late 1980s, when the yen was strong and domestic demand booming, and has continued into the 1990s. Kanzaki, Settsu, Daishowa and Jujo have set up new paper mills in the US and Settsu in Portugal and Spain, while Nippon and Marubeni are involved in a new pulp operation in Sumatra and 19 other Japanese firms are working to expand a huge pulp operation in Brazil. Japanese firms are also involved in paper operations in Indonesia, Thailand, Brazil, Malaysia, China and Singapore. In Canada, meanwhile, Daishowa, New Oji, Honshu, Hokuetsu, and Mitsubishi control gigantic 850,000 tonne-per-year pulp operations in Alberta, with concessions covering over 10 million hectares. As a result of such expansions, Japan's share of world wood pulp imports went up from less than five per cent in 1965 to nearly 14 per cent in 1989. In 1993 Japan imported about 3.3 million tonnes of pulp, 71 per cent from North America, two per cent from Brazil, Chile and Indonesia, and smaller amounts from New Zealand, the Nordic countries, Portugal and South Africa. Japanese firms have also begun buying paper from Canadian companies such as MacMillan Bloedel (*AP&P* 31 (2); Dargavel 1991; Olsson 1995; Whitham 1994).

Chapter 4

Impacts on People and their Environment

Nearly every human activity has some impact on society and its environment, positive, negative or both. Agricultural crops are no exception. While such crops are neither negative nor positive in themselves, their cultivation can be defined as basically positive if it turns out to be sustainable in the long term, if the process it begins is reversible, and if it benefits local people. On the other hand, it can be defined as fundamentally negative if it is not sustainable in the long term, if it sets in motion processes which are not reversible, or if it results in losses for local communities.

In recent years, eucalyptus, because it is planted so extensively, has become a symbol of large-scale tree crops in the South. However, it would be wrong for analyses of the impacts of such plantations to centre on the botanical or ecological characteristics of eucalyptus. The problem lies not in any particular species nor its unique biological features, but in how it is used. The issue would not be substantially different if any other tree — native or exotic — were planted on a large scale to supply industry. As shown in chapters 8 (Chile) and 10 (South Africa), problems generated by industrial pine plantations are very similar to those created by eucalyptus.

Unfair to tree plantations?

Many foresters and plantation owners claim that environmentalists have unfairly highlighted the defects of tree plantations while paying less attention to other crops that — according to them — do just as much to degrade the environment. Brazilian forester Walter de Paula Lima (1993) states, for example, that 'it is interesting to observe the duplicity normally encountered when comparisons are made between agricultural crops and forest trees'. The Uruguayan pulp firm Fábrica Nacional de Papel (1992) puts forward the same criticism, saying that 'it is worthwhile showing the moral double standards which seem to have been established to judge agricultural crops'. Assuming that this accusation is made in good faith, it requires several responses:

- It is false that environmentalists have been especially harsh in their treatment of tree crops. On the contrary, the environmental movement has traditionally concentrated on deforestation, indiscriminate use of toxic chemicals in agriculture, the disappearance of biodiversity, the impacts of large dams and nuclear power stations, and other themes, only recently turning to tree plantations. Indeed, many Northern environmentalists, far from being too harsh, still automatically assume that all tree plantations must be good.

• Where environmentalists have taken an interest in the problems of large-scale tree crops, this has been motivated largely by the complaints of local affected peoples. Where investigations have been carried out, they have tended to confirm the existence of deleterious effects, a fact which is increasingly accepted by governments and businesses alike.

• The Green Revolution — of which the type of forestry development discussed in this book is one part — has been denounced by many environmentalists on both ecological and social grounds. Criticism of tree crops in many ways merely follows on from criticism of the Green Revolution — which, of course, is mainly directed at non-tree crops. Over the years, environ-mentalists have thus paid at least as much attention to the problems of non-tree monocrops as they have to tree plantations.

• Large-scale industrial monocultures, by virtue of their extent alone, can generate large-scale problems, environmental and social. Given the worldwide promotion of such plantations, disseminating information about their potential impact should be an important priority.

The two libraries

Scientific research on tree plantations falls into two libraries. On the one hand, a variety of scientific work has been conducted to prove that monocultures of eucalyptus, pine and other species do not have large negative effects, some even maintaining that they have positive consequences for society and its environment. A great deal of equally important scientific work, on the other hand, has concluded that such plantations do substantial damage, both social and environmental.

Why this divergence of views? Is it that one or the other group of scientists is doing bad science? If so, in what sense? Or is it that they are all incapable? Or is it rather that we should not expect silviculture to be able to provide all the answers to questions about the biological or social effects of plantations? Several reasons may perhaps be given for the existence of the two libraries.

First, modern science is not an objective, monolithic construction located outside society. Any scientific work is coloured by the experience of the author and linked to her or his scale of values and vision of the world. Scientific research may also be reasonably expected to be often directly affected by scientists' material interests. Walter de Paula Lima (1993), following extensive research on the environmental impacts of eucalyptus, concluded that eucalyptus has no important negative effects on water, soils, flora or fauna. Although Lima's book is accompanied by the obligatory disclaimer that it does not take any 'stance of defence or attack, as fits true scientists, making an impartial and sensible analysis of the issue', it was in fact made possible only through the collaboration of Brazil's National Association of Pulp and Paper Producers, 'with valuable financial support' from almost 40 companies, including major powers in pulp and paper such as Aracruz, Bahia Sul, CENIBRA, Champion, Monte Dourado, Votorantim, Riocell, Norcell, Ripasa and Klabin. Khon Kaen University scientists in Northeast Thailand, similarly, found clear evidence that, even under artificially favourable conditions, the use of pulp mill wastewater for irrigation would damage plant growth severely and necessitate large investments in techniques to fertilize and remove salt from the irrigated soil. Yet the scientists concluded, contradicting the data they themselves had collected, that the use of such wastewater for irrigation was 'feasible'. The research was supported by the Phoenix Pulp and Paper Company, Ltd — a firm eager to find new ways of disposing of its water-borne effluents (Jirasak et al. 1993).

A second reason for the existence of the two libraries is that silviculture (which traditionally focused on wood production for the market) cannot by itself offer a comprehensive analysis of the impacts of plantations. The problem rests not with the discipline of silviculture specifically. Rather, it lies in the reductionist tendency of modern science generally to divide reality into isolated parts while insisting on identifying some of those parts with the whole (Shiva 1993). Any research on the impacts of plantations needs the participation of many agents to arrive at serious conclusions. Some knowledge forms part of the cultural heritage of local peoples. Other knowledge can be arrived at by external agents (environmentalists, natural and social scientists, forestry experts, and so on). Uncoerced interaction between both groups of agents is more likely to result in realistic, fruitful conclusions than 'scientific' studies conducted in isolation from the community.

Third, because pressures to establish plantations exist worldwide, monoculture tree crops are grown today in a wide variety of social, political, economic and environmental circumstances, and under many different management regimes. The results of studies of the effects of plantations on biodiversity or soils in a single location thus cannot automatically be generalized. The same is true of social impacts.

Finally, scientists studying plantations often do handle data in a remarkably unscientific fashion. Totally dissimilar circumstances, for instance, are often compared as if they were similar. It would seem to be a mark not only of scientific caution but also of common sense to assume that a native eucalyptus forest in Australia cannot be compared with a plantation of the same species outside its native habitat, that patches of trees planted in agroforestry systems are not the same as large-scale monocultures; and that an intensively-managed industrial plantation of fast-growing species will not have the same characteristics as an unmanaged and unexploited plantation. Nonetheless, all these comparisons are regularly made, and the conclusions presented as scientific evidence, with the result — whether intended or unintended — that the public becomes confused.

This chapter aims to avoid this error. It will not rely on studies of plantations on degraded soil, nor studies of small plantations, nor studies of agroforestry systems, nor studies of narrowly-focused laboratory interactions. Instead, it will be based on real-world observations of large-scale tree monocultures planted to supply industry. There is already evidence concerning this type of monoculture in enough different locations to draw some firm general conclusions, and it is these that constitute the core of this chapter. At the same time, however, the chapter will not pretend to deal with all of the impacts of plantations, nor claim to decide in advance which are the most important, since the effects of plantations of most importance to local communities vary from one place to another. The chapter aims, rather, at offering observations which may help clarify the issues.

Impacts on water

Tree crops have been publicised as performing functions similar to those of a forest in the maintenance of the water cycle. Some plantation proponents have even gone as far as to state that the tree plantations in prairie ecosystems help to regulate the water cycle (JICA 1987). Such affirmations are groundless.

Changes in the water cycle

The water cycle can be described as follows: part of the rainfall is intercepted by the vegetation and evaporates, while the rest falls to the ground and either runs off superficially or filters into the subsoil. Part of the water absorbed is used by plants and returned to the atmosphere; another reaches the underground water table and circulates internally toward water courses or springs and the rest evaporates (Shiva and Bandyopadhyay 1987).

In ecosystems which have not been radically modified recently, the naturally-occurring vegetation possesses characteristics which ensure long-term balance in the water cycle. The morphology and physiology of the component species of the local ecosystems tend to be adapted to make most efficient use of available rainfall. Large-scale tree plantations, however, modify all of the following:

- *The ratio between the amount of water intercepted by the foliage and the amount of water reaching the ground.* The foliage of a plantation differs from that of a natural forest, a savanna or a prairie, in biomass, height, form of cover, and shape and distribution of leaves and branches. Plantations also generally lack undergrowth. These characteristics change the quantity of water intercepted and evaporated. Thus, the soil will tend to receive either more or less water than that received under the original vegetation.

- *The ratio between the amount of water which runs off the surface and the amount of water which infiltrates the soil.* This is affected by factors such as the type of humus generated by the plantation and the quantity of accumulated leaf litter, which facilitate or complicate the absorption and infiltration of water which reaches the surface. The volume of water which crosses the canopy also affects this ratio. Soil compacted by heavy machinery, in addition, impedes infiltration, encouraging evaporation.

- *The ratio between the amount of water evapotranspired and the amount of water which infiltrates through to the subsoil water supply.* This ratio depends largely on how much water is used by the species planted. Growth rate is directly related to water consumption. In plantations that use the fastest-growing genotypes of fast-growing species, water consumption tends to be extremely high.

To begin with, then, we can be almost certain that a plantation will introduce changes in the water cycle. The type and degree of changes will not only depend on the species planted and its management, but will also be affected by the local climate (volume of rainfall, seasonal changes, droughts, temperature, winds), topography and soil type.

Hydrological deficits caused by plantations can result in a number of impacts such as:

- *Reduced availability of water for other livelihood and industrial activities.* In Espírito Santo, Brazil, for example, a number of water sources are alleged to have been affected by eucalyptus plantations, depriving local people of fish and reducing farm yields (*IAD* 6/7.1992). In the Tarawera River in New Zealand, lower flows resulting from big pine plantations 'are creating problems for downstream users, including ironically, the Tasman pulp and paper mill where toxic discharges have less water available for dilution' (Rosoman 1994).

- *Problems of water supply for hydroelectric generation systems,* such as that being faced by Electricorp in the MacKenzie basin of New Zealand (Rosoman 1994).

- *Discontinuity in the flow of watercourses in low periods.* In South Africa, for instance, during the growing season, flow frequently ceases in areas where plantations have replaced natural, non-forest vegetation such as shrub or bush (Wicht 1967, cited by Sawyer 1993).

- *Increased impacts of droughts.* In the climatically drought-prone zone of the coast of the Maule and Talca in Chile, for example, excessive water consumption by rapid-growth tree plantations has been noted to deplete the groundwater reserves (CODEFF 1994). Caroline Sargent (1992) states that 'where there are downstream water shortage problems, the excess water transpiration of fast growing plantations...is likely to be deleterious, reducing net flow and contributing to drought potential'.

- *Less water for local communities and urban centres.* In Chile, houses and agricultural plots have had to be abandoned for lack of water and the town of Angol began facing serious water supply problems eight years after plantations were established in the area (Cruz and Rivera 1983). As a general rule, where trees replace non-forested land uses, 'the overwhelming evidence from catchment research is that following reforestation, groundwater levels are lowered and stream yields are reduced, both effects being more pronounced during the dry season or growing season' (Hamilton and King 1983).

•   *The modification or destruction of other natural ecosystems, such as wetlands.* In Natal, South Africa, Porter (1990) points to industrial plantations as one of the principal threats to the St Lucia Wetland Park, and in northeast Thailand plantations have damaged fish spawning grounds in seasonally-flooded riverine environments.

Rejecting empirical observations

Faced with a lack of agreement between their theoretical models of plantation water use and some of their empirical observations, some foresters have chosen to disregard the empirical observations. In Chile, for example, plantations of *Pinus radiata* have in some cases caused springs and other natural sources of water to dry up, while at the same time rendering the flows in rivers irregular, with valleys being inundated during the rainy season. In the area of Concepción, the flooding of the river Andalíen has forced people to abandon most of their farms. In the same area, the river Mininco now floods roads and homes of local people (Cruz and Rivera 1983).

Faced with such observations, one soil professor and forestry expert stated that 'the pine was brought in as a high yield product and it needs sufficient water to produce wood. Nonetheless, I would doubt that a pine forest consumes a quantity of water much higher than a natural forest does'. He then contradicted himself by affirming that 'in terms of consumption, as the pine is a rapid growth species and has a greater biomass, it would be expected to have water consumption several times higher than the native forest' (Cruz and Rivera 1983).

Less confused were the simply-expressed observations of a Chilean farmer from Rere. Having inherited an area of eight hectares, nearly totally planted with pines, the farmer found that he could get no water even for household use. As the plantation matured it was sold, with all the pines being cut and only a small copse of oak in a ravine being conserved. To his surprise and that of his neighbours, a dried-up stream then reappeared (Cruz and Rivera 1983). Precisely parallel observations of the recovery of streams, wells, and *nong* — standing bodies of shallow water crucial for water buffalo and other livestock — have been made by northeastern Thai farmers following the harvest of eucalyptus plantations (PRED 1996).

A useful function for science, in such cases, would be not to deny that the disappearance of sources of water was a result of the plantation but rather to look for the mechanisms involved. For example, had the Chilean pines or the Thai eucalyptus used an excessive amount of water? Had water failed to trickle through to the subsoil in Chile because of the covering of needles? Was there excessive evaporation from the foliage or soil in either country?

A similar instance of scientific denial occurred in Uttar Pradesh, India, where Mahashweta Devi, an elderly forest guard, told of his experience: 'We felled mixed natural forests of this area and planted eucalyptus . . . Our handpumps have gone dry as the water table has gone down. We have committed a sin' (Shiva and Bandyopadhyay 1987). Instead of supporting investigations into such occurrences, Tewari, the President of the Indian Forest Research Institute, simply denied them, writing in a contribution to a special issue of *Indian Forester*:

> Of late in India a lot of controversy has arisen over the water consumption behaviour of *Eucalyptus* planted in forestry programmes in social forestry. It has been alleged that *Eucalyptus* plantation consumes large quantities of water to the extent that they deplete local water resources such as streams, wells, etc. This notion does not appear to be correct as no experimental data in support has so far been presented . . . There is no scientific basis in the popular fallacy that *Eucalyptus* lowers the ground water table (Shiva and Bandyopadhyay 1987).

In Spain, similarly, 'experts' have sometimes ignored the testimony of farmers such as Constancio Romero, from Aroche, who noted that

> on this farm there were irrigated crops and livestock, but once the Eucalyptus was planted in part of the hills, the waters reduced and we couldn't keep working here . . . With the Eucalyptus everything was left devastated: no grass, no animals, no nothing. There is a lot of land in the lower part of Tariquejo which was left without water. . . . It is very sad that people with animals find themselves without water because of the Eucalyptus (PSOE 1979).

In Thailand, meanwhile,

> many plantations are situated on water tables and ground water sources of local villages, especially in the dry Northeast. The eucalyptus tree absorbs and reduces the ground water so much that the villagers cannot irrigate the rice fields. This environmental damage is greatly resented by many farmers, and is one of their main reasons for complaining against the eucalyptus reforestation policy (Van Ginneken 1993).

Yet instead of taking such views seriously, technicians have sometimes promoted eucalyptus trees in the Northeast 'precisely because of their high water uptake. By lowering the water table, they [are held to] reduce the risk of salinity in paddy fields' (Van Ginneken 1993).

Accepting reality

Forestry experts, of course, are not always in the position of having to discount empirical observations. In New Zealand, the electricity company Electricorp faced up to the possibility that plantations of *Pinus radiata* had reduced the flow of water into reservoirs behind hydroelectric dams, thus threatening power generation. Research showed that a possible reduction of 25–30 per cent in water levels would result if pine plantations were substituted for local grasslands. This was not mere speculation. In the MacKenzie basin, where tree crops had been planted (pine trees now covering 31 per cent of the catchment), flow reduction of the Tarawera river has been calculated at 27 per cent for the period 1964–1992, and in Moutere Catchment, Nelson, 55 per cent of surface runoff and 70 per cent of the water feeding the underground water table had disappeared (Rosoman 1994).

South Africa is perhaps the country where research into the effects of industrial plantations on water regimes has been carried out over the longest time. Technicians there agree wholeheartedly with farmers that large plantations are voracious water consumers. As one South African forester affirms, 'commercial timber plantings use larger quantities of water than shorter vegetation types such as scrub, herbs and grass', concluding that 'afforestation tended to deplete substantially both the total annual water yield and the base flow in the dry season' (Le Roux 1990). As a result, in 1972, the Forest and Water Department established a limit for the area that could be covered by tree plantations in each water catchment area or subdivision thereof. Plantations were permitted to reduce the surface water runoff by no more than ten per cent. Even so, there is still concern that 'the indiscriminate planting of trees may seriously affect the viability of the springs and wetlands in many catchments' (Le Roux 1990).

Finally, even the United Nations' Food and Agriculture Organization (FAO), which has been — and still is — one of the main promoters of monoculture tree crops, is beginning to accept that the 'plantation of extensive areas of eucalyptus in any deforested water catchment area substantially reduces the water production of this source, and the felling of the trees will increase it':

> The effect of the eucalyptus on the reduction and production of water is probably less than that of pine and more than that of other hardwoods; but all tree species reduce the production of water to a greater proportion than scrub or grass. Consequently, when the water production of a water catchment area or the state of the water table in the low-lying adjacent territories is affected,

the situation must be considered very seriously before large scale planting or felling projects are carried out (Poore and Fries 1985).

Conclusions

Monoculture tree crops have had negative effects on water cycles in widely differing areas. Although this does not imply that such negative effects will occur in and around all plantations, it is sufficient to justify obligatory environmental impact assessments wherever such plantations are planned. The problem, again, is not with any particular tree species, but with high-yield industrial production, which tends to consume quantities of water in direct proportion to rapid growth. The large scale of industrial plantations, in addition, means that the problem can affect entire river basins, with serious repercussions on the economy, society and the environment.

Impacts on biodiversity

According to the international Biodiversity Convention signed at the Rio de Janeiro 1992 Earth Summit, biodiversity includes 'diversity within species, between species and of ecosystems'. All of these types of biodiversity are threatened by industrial plantations — which are often supported by the same governments which signed the convention.

Human activities affect biological diversity in many ways, whether by substituting one ecosystem for another, exterminating or decreasing the numbers of certain species, consciously or unconsciously encouraging the explosive development of others, or helping certain characteristics predominate within one species. One of the main ways in which human activities reduce biodiversity is through the deliberate replacement of diverse ecosystems by homogenous ones through agriculture, animal husbandry, fishing and forestry.

Tree monocultures and biodiversity

Any industrial plantation affects not only ecosystems, but also local species and genetic variety. These plantations are constituted by a dominant species (normally exotic), which grows in extensive homogenous blocks under intensive management. This management involves soil preparation, fertilization, herbicide use, clearing of undergrowth, elimination of diseased trees, thinning, persecution of animals which can damage the trees, and periodic logging. All these factors cause profound changes in the flora and fauna of local ecosystems, which can affect the region as a whole.

Industrial plantations may replace crops, prairies, forests or any other ecosystem. Given their scale, they generally include both areas in which there has been a great deal of production-related human interference and areas in which there has been relatively little.

In many cases the biggest impact of plantations on biodiversity is made before a single tree is planted. In many countries, native forests are destroyed — legally or illegally — partly in order that they may be replaced with plantations (see chapters 7, 8, 11 and 12). Postel and Heise (1988, cited by Sawyer 1993) have calculated that at least 15 per cent of all plantations in tropical countries were established at the cost of natural closed forests. Native hardwoods themselves have been used, or are slated to be used, as pulp raw material in Papua New Guinea, Cameroon, Indonesia, Malaysia, Nigeria, Chile and other countries (Dudley, Stolton and Jeanrenaud 1995).

Biodiversity losses in such cases are enormous. In 1993, for instance, Veracruz Florestal was accused of cutting down Brazilian Mata Atlantica forests to set up eucalyptus plantations. The Mata Atlantica holds a record in biodiversity; a recent survey found 450 species of trees in only one hectare of forest (Faillace and Miranda 1993). In densely-settled areas, especially in the tropics, big plantations also frequently displace large numbers of farmers. Such migrants are often forced to clear areas of natural forest elsewhere to establish new farms (see the case studies of Part Two, especially chapter 12). In this case, plantations have a powerful indirect impact on biodiversity.

The changes plantations bring about in soil and water cycles also affect plants and animals. Such changes occur preponderantly within plantation boundaries. However, changes in water supply can have far broader impacts as well. The high water consumption of plantation trees can severely modify adjoining wetlands, or even cause them to dry up, along with local watercourses. This results in the disappearance of, or reductions in, plants and animals dependent on these habitats. Even when these extremes are not reached, changes in hydrological systems brought about by plantations can affect any species dependent on local watercourses.

Impacts on plants

Industrial plantations begin with preparation of the soil. Most local plant species are removed. Pioneer species which return are destroyed either by mechanical clearing or by herbicides. Once the plantation trees attain a certain age, they impede the development of most other plant species as a result of increased shade, accumulation of leaf litter and dead branches on the ground, competition for water and nutrients, the cumulative effects of certain changes in the soil, and the allelopathic effects of some species which produce chemical substances that negatively affect the growth of other species (Shiva and Bandyopadhyay 1987, Rosoman 1994, Barnett and Juniper 1992).

The few species which do manage to survive in the plantation or in fire-breaks are periodically eliminated to reduce the risk of fire. The ecological characteristics of the plantation species themselves, together with the intensive management and felling carried out when the trees reach the appropriate size for processing, ensure that various flora which might otherwise be associated with such trees (epiphytes, parasites, climbing plants, and so on) cannot develop. As will be suggested below, the impact of this reduction of plant biodiversity on local livelihood can be severe. In a single community in Yasothorn province in northeast Thailand, for example, eucalyptus' destruction of local grass cover deprived local cattle and water buffalo of grazing, forcing a dozen families to abandon their homes (PRED 1996).

Photographs are sometimes shown of plantations with extensive undergrowth (an example is in Lima 1993). Such cases, however, are typically not industrial monocultures, but abandoned or unmanaged plantations in which local pioneers have begun to move in, occupying, in particular, spaces which open up between trees in old plantations.

It is sometimes claimed, too, that in some cases plantations increase local plant growth, as for example when they replace non-forest communities. Two FAO consultants, for example, claim that plantations 'create a forest environment which normally produces a beneficial result'. They admit, however, that it is 'improbable' that plantations 'encourage the species characteristic of the previous unpopulated terrain' (Poore and Fries 1985), and another FAO study concedes that 'when indigenous plant communities (forest, woodland or grassland) are converted to monospecific or polyspecific plantations of native or exotic species, with the main purpose of wood production, generally there will be a reduction in both habitat and species diversity at that site' (FAO 1992).

Some of the governments and companies involved in large-scale industrial tree plantations have been forced, in response to public pressures, to adopt the opposition's discourse and to embellish the monotonous plantation landscape with some patches of native forest or plantings of native trees. These changes, however, are necessarily merely cosmetic, since the main purpose of the plantations — to provide huge amounts of uniform industrial raw material — remains unaltered. As Philip Fearnside of Brazil's National Institute for Research in the Amazon remarks, modifications which bring in a 'mix of a few species, including some that are not exotic, do not substantially change the impact of very large-scale plantations from the standpoint of biodiversity' (Fearnside 1993). Yet such changes do sometimes serve the purpose of confusing the public and dividing the opposition. The Brazilian case (see chapter 7) is particularly revealing in this respect.

Plantations have an ecological impact on flora outside plantation boundaries as well:

- Some species commonly used in plantations, when put into suitable environments, reproduce beyond the plantation and become pests to the local vegetation. Such biological pollution occurs in some cases on a massive scale, as with *Pinus patula* and *Acacia melanoxylon* in South Africa (Bainbridge 1990, Rosoman 1994) and *Pinus pinaster* in Uruguay (Carrere 1994).

- Some diseases or pests that did not exist in the plantation region may begin to affect native vegetation. In India, a fungus which developed on an exotic pine (*Pinus radiata*) is now threatening the survival of the native pines *P. roxburghii* and *P. wallichiana*. In Kenya and Malawi, an aphid which began by attacking the exotic cypress *Cupressus lusitanica* moved on to Malawi's national tree (*Widdringtonia nodifolia*) and another local tree (*Juniperus procera*) (Barnett and Juniper 1992).

- Fires originating in or linked to plantations can seriously affect flora over immense areas. As Cavieres and Lara (1983) note, the *copihue*, Chile's national flower, has been seriously reduced in one study area of that country as a result of such fires, 'with only a few scarce weedy examples now existing'. In northern Thailand, too, fire used to simplify the structure of plantations has entered neighbouring areas of forest and threatened non-tree plant growth used by local villagers for various purposes.

- Fertilizers, herbicides and pesticides carried by wind or water may have impacts far beyond the plantation area. These chemicals contaminate soil, waterways and the atmosphere, and affect people, plants and wildlife (Rosoman 1994). In Brazil, the inhabitants of the fishing community of Caravelas went to the courts to demand an investigation into a recent reduction in crabs and other species which they linked to the use of agrochemicals by Bahia Sul's plantations (CEPEDES/CDDH 1992).

Impacts on animals

For most local animals, a plantation is a desert, lacking food, shelter and opportunities for reproduction. The species commonly used in plantations are exotic, and their principal advantage stems from the near absence of 'pests and diseases' in the new environment at the time they are planted. Yet however positive this may be for the forest investor, it is not so for the local fauna whose habitat is replaced.

For some species, plantations provide shelter from predators, but this can lead to a drastic reduction in the predator population and uncontrolled increases in the prey population. In many regions of Chile, for instance, plantations have caused a drastic reduction in fox numbers and a related increase in numbers of rodents and rabbits, which in turn affect the pines in the plantations (Schlatter and Murúa 1992).

Imbalances generated by plantations affect a very wide group of species, the majority of which are unobserved by non-residents and non-specialists. The enormous variety of life forms existing within the soil (including bacteria, viruses, fungi, small insects, and worms) can suffer large impacts from the combination of changes in leaf litter and other decomposing vegetable matter and changes in the chemical composition and structure of the soil. The use of agrochemicals also importantly alters soil flora and fauna (Rosoman 1994).

Industrial plantations are characterised by intensive management, based fundamentally on calculations of economic yields. Trees never reach full maturity, but are felled when they attain optimum sales dimensions or their growth slows. Plantations thus harbour few of the mature, diseased or dead trees which serve as microhabitats for fungi and insects, which in turn serve as food for other animals. Epiphytes and climbing plants which support other fauna also tend to disappear (Barnett and Juniper 1992).

A small group of species manages to adapt to newly-created plantation environments. Even some of these species, however, are exterminated because they impede plantation development. In Uruguay, Argentina and Brazil, for example, leaf-cutting ants — one of the few insects which can feed on pine and eucalyptus plantations — have to be poisoned. The contamination which results, of course, can affect other animals which are inoffensive to the plantation trees. The few species which manage to adapt themselves to plantation ecosystems, moreover, suffer the destruction of their new habitat every few years when harvest time comes round. In the case of eucalyptus this happens every six to ten years, and with pine, every 12 to 20.

The impact of a plantation on animals, like that on plants, goes beyond its boundary, as species benefiting from the plantation increase in number and those harmed by it decrease. Fires beginning in plantations spread into surrounding ecosystems, while agrochemicals 'affect aquatic flora and fauna within and outside plantations when waterways become polluted with . . . minerals or chemicals' (Sawyer 1993) (see examples in chapters 7, 8 and 10).

How imbalances affect plantations themselves

The homogeneity of extensive tree plantations constitutes a serious problem for the plantations themselves. The great initial advantage of exotic trees —the absence of local fauna accustomed to using them as food — can become an Achilles heel in the long term, when predators adapted to this species do begin to appear. At that point the food desert becomes a feast for one species, which can expand exponentially and seriously damage or annihilate whole plantations. Such was the case in Uruguay with *Pinus radiata*, which had to be abandoned due to serious attacks by the pine shoot moth *Ryacionia buoliana*. A similar fate befell *Gmelina arborea* in Brazil, and monoculture tree plantations established by the Paper Industry Corporation of the Philippines (PICOP) have been plagued by pests for the same reason.

Chemicals developed for agriculture, moreover, are often awkward to use on plantations, particularly once the trees have canopied, and can contaminate wide regions. It is extremely difficult to apply chemicals effectively to dense stands of trees, and if airplane spraying is used, the amount of pesticide needed to guarantee coverage is enormous. These chemicals, sprayed from well above the level of the trees themselves, are necessarily partly carried away by the wind to contaminate large areas outside the plantation itself. In New Zealand,

> more than 30 brands of herbicide, pesticide and fungicide are used on tree plantations . . . , including highly toxic and persistent organochlorines. Large areas are sprayed with different chemicals. For example, around 10 per cent of plantations are sprayed on average 3.5 times for *Dithistroma* control. This amounts to about 90,000 hectares sprayed every year over the past 14 (Rosoman 1994).

Even though other methods of control are being developed in silviculture — for example, density control (controlling the number of trees per hectare to make the plantation less hospitable to certain pests or diseases) or biological control (introducing a predator of the pest population) — it is certain that the more monoculture pulpwood plantations spread, the greater the risks will be (Davidson 1987, cited by Sawyer 1993). Numerous pests now affect eucalyptus plantations in Brazil, whereas almost none were reported early this century. Second and third rotation pine plantations in South Africa are also suffering new infestations (Evans 1986 and Ball 1992, cited by Sawyer 1993).

Conclusions

By definition, forestry development based on monocultures affects biodiversity. The internal logic of the monoculture plantation concept — carried to extremes by the Green Revolution in agricultural crops, which influenced later developments in forestry, fishery and dairy farming — implies the substitution of the diversity present in nature with the homogeneity of the industrial process. In forestry, today's large-scale plantations are the paradigmatic expression of this concept. There, genes, seeds and plants alike are controlled for industrial gain; the recycling of nutrients is replaced by the adding of fertilizers; competition is removed by the use of herbicides; and growth is controlled by spacing, thinning out, and so on. The contemporary plantation has been defined as the 'roofless factory': at one end raw materials (genes) are introduced, other elements and energy are applied along the way, and out the other end comes a homogenous product fulfilling predefined conditions.

The extent to which this process undermines biodiversity depends on plantation species, scale, and management methods — but that it has a negative impact on biodiversity can hardly be questioned. On a global scale, biodiversity cannot be conserved by attempting to fence off a few untouched areas in the middle of an increasing sea of homogeneity. As Vandana Shiva puts it, 'not until diversity is made the logic of production can diversity be conserved' (Shiva 1993).

Impacts on the soil

Many existing studies of the impacts of industrial tree plantations on soils confuse the issue by citing irrelevant research. For example, a recent FAO publication notes that the

> effects of uncropped eucalypts on soil quality have been compared with [those of] other species and . . . treeless areas. The studies were mostly in India and the Mediterranean and are fairly recent. Eucalypts were found to have a beneficial effect on soil structure and compared favourably with pine and *Shorea robusta* (*sal*, a local tree). On treeless sites eucalypts improved soil fertility through decayed leaves and litter (FAO 1990).

This quotation is used to suggest that, in general, eucalypts improve soil quality. Yet the example refers to unharvested plantations, while large monospecific plantations of any species are normally planted to be harvested and not to improve the soils. Moreover, in the real world in which investors are bent on obtaining high yields, large-scale plantations tend not to be established on degraded soils of the kind the FAO appears to describe, where trees grow poorly. When discussing the effect of real-world commercial plantations, it is more relevant to examine a later passage in the same work, which says that, in managed and harvested plantations, the 'nutrient capital changes considerably because nutrients are removed from the site'.

Once such confusions are cleared up and the discussion is concentrated on industrial monocultures, plantation proponents are forced into a final argument, which relies on a comparison between industrial tree crops and the agricultural crops of the Green Revolution model. Industrial plantations, the argument goes, should not be the subject of special environmentalist concern since they are much less degrading to the soil than such crops. For example, the FAO cites a study showing that 'the amount of nitrogen taken in by the cereal crop is two and a half times more that the amount taken by the eucalypt plantation, and 15 times more in the case of phosphorus' (FAO 1990). Green Revolution-style silviculture, in other words, can defend itself only by saying that it is not quite as bad as the movement from which it draws much of its inspiration. This defense, in addition to being ineffective against critics of the Green Revolution in agriculture, implicitly abandons the premise that one of the main points of tree cultivation is to foster *non*-agricultural, forest-like environments.

The nutrient cycle

Trees obtain nutrients needed for growth from the soil. As Rosoman (1994) explains, in natural ecosystems many of the same nutrients are used again and again in a relatively closed cycle. Tree roots draw out minerals dissolved in water from the soil and carry them to the leaves, where they are transformed into organic material and used for the tree's vital functions. Leaves, branches, flowers and so on then fall to the ground, where various organisms decompose them and liberate minerals that can then be taken up again by the roots. When the tree itself dies and decomposes, more nutrients flow into the cycle.

Even cycles which are relatively closed in this respect, however, have some inputs and outputs of nutrients. Inputs come mainly from the atmosphere (as salts or other materials deposited on the leaves and which reach the soil with rainwater), from the decomposition of the rocks from which the soil originates, from watercourses (particularly where floods occur), and from the droppings and decomposing bodies of animals. Outputs leave the system through wind and water erosion, through the percolation of dissolved nutrients to the underground water table or to layers of the soil inaccessible to plants, and through animals which extract organic matter from the system and deposit it outside the area.

Nitrogen, in addition, can enter the system through the action of certain bacteria present in the roots of some plants or through rainfall. It can leave the system, meanwhile, through oxidation of organic material or through processes which liberate it in a gaseous state.

Trees need some nutrients in relatively high quantities, while they use only a little of others. The former are called macronutrients and consist principally of nitrogen, phosphorus, potassium, calcium and magnesium. The latter are known as micronutrients (boron, copper, zinc and others) which, though not required in great quantities, are just as indispensable for tree growth.

Plantations and the soil

In ecosystems little affected by human (especially industrial) interference, nutrient cycles tend to be in relative balance between incomings and outgoings. Not so for intensive monoculture tree plantations. Moreover, plantations with fast-growing species and rapid-rotation felling bring about much more important modifications in the soil than do unmanaged or unexploited plantations.

The direct impacts of such plantations on soils derive from the presence of the trees themselves, and include changes in the recycling of nutrients and in chemical and physical soil composition.

In the majority of commercial plantations, an imbalance arises between the nutrients taken up by the roots and those given back to the system by dead organic matter. Because trees such as eucalyptus and pine tend to reduce the action of decomposing agents such as fungi and bacteria, nutrients contained in the leaf litter are not freed up in a form which would allow them to be taken up easily by roots. Chemical changes such as the acidification of the soil and the introduction of new chemical compounds make life more difficult for many decomposers, and changes in humidity, temperature, and light have an additional impact. The leaf litter of such pulpwood trees themselves contains tannin, lignin, oils, waxes, and other substances which are difficult to digest or even toxic for soil flora and fauna. Many decomposers not able to adapt simply disappear. As a result, the leaf litter decomposes only slowly, accumulating on the soil. One study in Nigeria showed that while leaves from native forests decomposed in two to seven months, leaves from introduced plantation pines took three to six years to do the same (Barnett and Juniper 1992).

In non-commercial plantations, this problem is less serious, as a balance is eventually reached. Nutrients are not exported, litter eventually decomposes and the plantation as a whole stops growing. In some cases, indeed, non-commercial plantations can help enrich soils by reintroducing nutrients which had previously been located in deeper layers of the soil where the roots of native trees, shrubs or grass did not reach. In commercial plantations, however, the soil becomes poorer in direct relation to growth rates and felling rotations. Fast growth combined with slow litter decomposition implies that trees are extracting nutrients faster than they are replacing them.

Tree plantations can also lead to greater acidification of the soil and to changes in its physical properties. According to Rosoman (1994), acidification is produced by a combination of two factors. One is the reduced pace of decomposition of organic matter which is characteristic of plantations. The other is the export of nutrients. Trees take negatively charged ions (cations) from the soil, while leaving positively charged ions such as those of hydrogen and aluminium. In ecosystems subject to little industrial interference, cations are eventually returned to the soil when the trees die and decompose, while in industrial plantations many of them are removed at the moment of harvest. Other cations remain locked in non-decomposed organic matter.

Impacts of planting and management

Industrial plantations begin with large-scale preparation of the soil. This adds to the danger of erosion, especially in areas with pronounced slopes. After the trees are planted, weeding is carried out manually, mechanically or with herbicides, to prevent other plants from competing with them. This destroys some or all of the native vegetation which has survived the initial planting. Thus the soil is left unprotected from erosion for a relatively prolonged period —sometimes as much as two or three years. Erosion can attack not only the surface soils of the plantations, but also areas where runoff collects. This has been the case in, for example, Galicia, where terracing for planting eucalyptus has resulted in serious erosion (Ruiz 1990).

When plantation trees are harvested, moreover, more erosion results, together with the wholesale export of nutrients from the plantation site. Three methods exist for harvesting plantations: (1) extracting whole trees, (2) extracting trunks together with their bark, and (3) extracting stripped trunks only. According to the method adopted, the export of nutrients from the system will be greater or lesser, though export occurs in all cases. In a study carried out in Brazil, the nutrient content of various components of a four-year-old *Eucalyptus saligna* plantation with 38 tonnes per hectare of aerial biomass was analysed as in Table 4.1. When only trunks were taken, nearly half of the phosphorus contained in the trees was removed, along with almost a quarter of the potassium and smaller percentages of the magnesium, nitrogen and calcium. When trunks were removed together with their bark, the export of phosphorus increased to 58 per cent, the magnesium to 44 per cent, the potassium to 39 per cent, calcium to 35 per cent and nitrogen to 20 per cent. Removal of the entire tree, of course, was the worst option.

TABLE 4.1
Nutrient content of *Eucalyptus saligna* plantation (I)

Component	Per cent of tree's total quota of elements				
	N	P	K	Ca	Mg
Trunk	12	49	24	8	14
Bark	8	9	15	27	30
Branches	17	14	26	34	17
Leaves	63	28	35	31	39
TOTAL TREE	100	100	100	100	100

Source: Poore and Fries 1985.

If these percentages are converted into kilogrammes per hectare, the figures in Table 4.2 result. At a normal commercial harvest age, even more nutrients — between four and five times the amount shown in the table — would be removed from the soil. Even without whole-tree harvesting, it has been estimated, three pine rotations on infertile soil will remove as much phosphorus from the soil as 20,000 years of natural processes (Adams 1978). (See also Holt and Spain [1986] and Jordan [1985] on depletion of soil carbon and nitrogen in *Araucaria cunninghammii* and *Gmelina arborea* plantations.)

TABLE 4.2
Nutrient content of *Eucalyptus saligna* plantation (II)

Camponent	Element				
	N	P	K	Ca	Mg
Trunk	21	12	42	17	6
Bark	13	2	26	57	13
Branches	30	3	45	69	7
Leaves	107	7	61	64	17
TOTAL TREE	171	24	174	207	43

Source: Poore and Fries 1985.

In sum, the higher the rate of growth, the higher the rate of extraction. The greater the amount of exported biomass, moreover, the more rapidly existing nutrients are exhausted, particularly when whole trees are harvested. This depletion of nutrients entails either that the plantation must be abandoned at some stage, leaving impoverished soil behind, or that chemical fertilizers must be applied. This second option is the one promoted by modern silviculture. Experience with agricultural crops, however, proves not only that chemical fertilizers do not provide a long-term solution, but that they also have other negative effects such as contamination of the above-ground and underground water supplies and impoverishment of soil microflora and fauna (Rosoman 1994).

This nutrient depletion occurs with both eucalyptus and pine, as is admitted even by studies done for active promoters of industrial plantations such as FAO and Shell:

> The short-rotation harvesting of eucalyptus, especially when the whole tree is used, leads to the rapid exhaustion of the reserves of nutritive elements in the soil. The above is a direct consequence of their rapid growth ... Certain evidence exists to show a greater removal of nutrients in pine plantations under similar conditions (Poore and Fries 1985).

> [W]hole-tree harvesting and short-rotation forestry does remove much of th[e] pool of nutrients, not only reducing soil fertility, . . . but also acidifying the soil (Good, Lawson and Stevens 1993).

Short-rotation plantations, in addition, require more frequent management interventions, which make the soil more prone to erosion and other forms of nutrient loss. Heavy machinery compacts the soil, making it difficult for water to infiltrate, also promoting erosion. Log extraction, meanwhile, breaks the soil's surface, leaving it exposed to the erosive action of rain. The growing tendency towards increased mechanization, and the replacement of chainsaws by large harvesting machines, is likely only to intensify damage to soils.

It is therefore absurd to suggest without qualification that any sort of tree planting protects or improves soil quality. All evidence shows that, on the contrary, industrial plantations degrade soils, and that their functions can in no way be compared with those of natural forests.

Industrial pollution

The type of plantations analysed in this book are geared to the modern pulp and paper industry, which has historically been one of the most contaminating industries, emitting 'some of the most toxic effluent that any industry can produce' (Kroesa 1990). Chemical pulping involves the use of sulphur-based chemicals, whose recovery gives the mills — and their surroundings — the smell of rotten eggs. Kraft pulping releases sulphur dioxide — a major contributor to acid rain — to the air at a rate of one to three kilogrammes per tonne of pulp, while sulphite pulping emits some five kilogrammes per tonne. Aluminium salts used in kraft pulping are highly toxic to certain fish and 'accidental spills, which occur frequently, can have disastrous effects on aquatic life downstream' (Kroesa 1990). Mechanical and chemi-thermo-mechanical pulping meanwhile results in the release of organic sulphur compounds, which, together with resin acids and other wood wastes, come to make up a highly toxic effluent, very difficult to degrade, and dangerous to fish.

The lignin contained in pulp gives it a brown colour, which in conventional bleaching is removed through the use of chlorine gas. The pulp is further whitened using chlorine dioxide or hypochlorite. On average, between 50 and 80 kilogrammes of chlorine is used to produce every tonne of conventionally-bleached kraft pulp. About ten per cent of this chlorine ends up combined with organic molecules from the wood and is discharged in the effluent from the mill. This produces toxic chlorine compounds called organochlorines, which tend to go directly into lakes, rivers and oceans. Extremely stable chemically, such compounds may be spread hundreds of kilometres from a single pulp mill, and are likely to accumulate in particularly dangerous amounts in animals high in the food chain.

Among these chemicals, the chlorinated ethers known as dioxins are among the most potent toxics known. According to the US Environmental Protection Agency (EPA), people regularly eating fish caught near pulp mills have 1,000 times more chance of developing certain cancers than control groups. In addition to being carcinogenic, the EPA has found, dioxins may have adverse effects on development, reproduction and the immune system in humans at levels close to those to which millions of people are already exposed. Tests conducted on animals show that dioxins can cause severe birth defects, stillbirths, sterility, and the feminization of males and masculinization of females (*ES&T* 29 (1), O'Brien 1990, Hocking 1991, Floegel 1994, Greenpeace International 1994, Kroesa 1990).

Pulp and paper mills also create a variety of other environmental and health problems. In the US, the EPA has reported, the paper industry is the third largest source of toxic pollutants. Mills generally release chloroform, carbon tetrachloride, hydrogen sulphide and sulphur dioxide into the air as well as organic residues and aluminum and other mineral salts into the water. In Webuye, Kenya, the Pan African Paper Mills' air and water pollution is believed to be responsible for a number of health problems; more than 60 per cent of the children born in Webuye during the last 15 years since the mill began operations in 1974 have had breathing problems between the ages of one to five. Effluents from the mill have contaminated the Nzoia River, affecting people who used to earn a living through small-scale fishing (Ong'wen 1994). The pattern is very much the same elsewhere (see chapters 7, 11 and 12). Research in Canada and the Nordic countries has meanwhile documented a staggering variety of fish disorders near mills, including skeletal deformities, reproductive problems, gill erosion, and deformed embryos (Dudley et al. 1995).

Although, due to public pressure, some factories and companies have recently made strides in pollution control, by, for example, reducing or eliminating the use of chlorine, the industry as a whole is still far from meeting reasonable safety standards. Worse, it is moving much pulp manufacture — where most pollution occurs — to the South, where controls are looser and production cheaper. In Germany 20 per cent of the installation costs of a pulp plant need to be earmarked for environmental protection, but Brazil's Bahia Sul Celulose set aside only 5.6 per cent for the purpose (CEPEDES/CDDH 1992).

Other socioeconomic impacts

Most large-scale commercial forestry plantations in the South are promoted and established in inhabited locations by government agencies, national and foreign businesses, multilateral banks, or other organizations external to the area. Although their aim is not to improve local quality of life, but to obtain large amounts of timber in the shortest possible time, both businesses and governments usually try to publicize locally the advantages that plantations supposedly will bring to local people. On a national level, too, it is often claimed that plantations and pulp mills will bring indispensible social and economic benefits, including direct and indirect employment, increases in exports, and support for national development.

Experience demonstrates, however, that the environmental problems of large-scale industrial plantations tend to be social and economic as well, and on both local and national levels.

Local effects

Plantations normally replace crops, grasslands, or old-growth, secondary, or scrub forests. Due to commercial necessities, they are rarely established on degraded soil, as their objective is short cycles of rapid growth requiring a certain level of fertility and water supply (see chapter 6) (Bazett 1993). Hence they typically occupy areas already being used in various ways by local people.

In some areas, the population is sparse and land tenure is both clearly defined legally and little contested. In other areas, where the population is dense and many landholdings are undocumented, local people's farms may be threatened when the state cedes land to forestry companies. In still others, plantations may usurp lands traditionally used by the community as a whole. These lands can include both communal fields and pastures, whose disappearance can force local people into overexploiting adjacent lands or forests (Lohmann 1991).

Large-scale, fast-growing tree plantations threaten local agriculture in less direct ways, too. They may, for instance, usurp water needed by other crops or by livestock. In South Africa, the Natal Agricultural Union is concerned that 'large scale afforestation of river basins is having a detrimental effect on the hydrological cycle of many of Natal's rivers and is creating hardships for riparian farmers downstream' (Fourie 1990), and similar concerns have been expressed in Chile (Cruz and Rivera 1983), Brazil (CEPEDES/CDDH 1992), Spain (PSOE 1979) and many other countries as well.

Species whose numbers had previously remained small, meanwhile, can rapidly become economic pests when large monocultural plantations are introduced. Such pests, which range from mammals, birds and insects to fungi and viruses, can affect both the plantations themselves and neighbouring agricultural crops and even livestock. In Uruguay, for instance, plantations have benefitted populations of parrots (*Myiopsitta monachus*), foxes (*Pseudolopex gimnocerus*) and the introduced wild boar (*Sus scrofa*), all of which can affect crops, poultry and sheep (Panario et al. 1991, Carrere pers. obs.).

Finally, the roots of plantation trees, especially eucalyptus, because they extend several metres horizontally, can also threaten neighbouring crops by competing for their water and nutrients. In northeast Thailand, villagers say that *Eucalyptus camaldulensis* is 'selfish' in its nutrient use (Lohmann 1991). Acknowledging this fact, some countries, such as Uruguay, have enacted laws requiring that a plantation's outermost line of trees must be at a certain distance from neighbouring land (Carrere 1993). Fast-growing trees, of course, can also cut off sunlight to crops planted in or near plantations. All these impacts are especially serious in densely-populated rural areas, where a reduction in production, however small, may have catastrophic effects, both threatening subsistence and raising food prices.

Plantations' takeover of forests can also lead to severe social, economic and cultural problems. Forests often supply water and compost for crops, fodder for livestock, and vegetables, game, honey, fruit, mushrooms, fibre, firewood, building wood, and medicine to local communities, and can also be a source of spiritual values. Where they disappear, diets, health, housing and incomes alike may suffer.

Where local people's land or forests are directly contested, they have reacted in a variety of ways. In Thailand, for example, farmers have petitioned responsible government officials, publicized their grievances through the press, mobilized marches on government offices, set up roadblocks, felled trees within plantations and even burned entire nurseries (see chapter 12). Repression has often resulted, with death threats, arson, and false arrests common.

Plantations often also create conflicts within local societies between those who oppose and those who assimilate to them, or, to use the familiar official rhetoric, between 'backwardness' and 'progress'. The construction of associated pulp mills, in addition, can burden local communities with thousands of migrants seeking work. The enormous economic clout wielded by large pulp and plantation firms meanwhile tends to distort local politics. As whole regions become almost totally dependent on the industry, local and regional governments are forced to bend their policies to suit its needs (see chapter 7).

In some social contexts, large-scale industrial plantations can create local employment, and this is one of the main arguments wielded everywhere both by state and corporations to convince local communities to accept the projects. However, 'very often plantation development results in a long-term net loss of employment' (Morrison and Bass 1992). Although figures vary widely from place to place and source to source, on the whole there appears to be agreement that industrial plantations cannot employ as many people as conventional agriculture, particularly family agriculture. Cases in which large-scale plantations have generated more employment than was already locally available, as in Uruguay, can be counted on the fingers of one hand.

The jobs created, moreover, are mainly for seasonal casual labourers, in particular during the plantation phase. Few climates allow planting to be carried out year round. On the whole, working conditions vary from bad to terrible (see chapters 7–10).

National-level impacts

The local social impacts of tree plantations, when aggregated, can give rise to national-level problems. For instance, the displacement of thousands of people by big plantation schemes — imposed or voluntary, direct or indirect — can swell shanty towns in the big cities of the

South, giving rise to increases in poverty, crime and prostitution and leading to land disputes with other communities. In the most extreme cases, as in South Africa, such dispossession can lead to violent inter-ethnic confrontations (Albertyn 1994).

The agroexport development model on which large-scale tree plantations in the South are usually based can also create economic problems on a national scale. One problem is concentration of wealth. Occupying large areas of fertile land, industrial plantations require state support and heavy, long-term investments varying from 600 to several thousand US dollars per hectare. In the vast majority of cases, they need tax exemptions, soft loans from foreign creditors, forestry research, road construction, improved port installations, and other subsidies which are extracted from a nation's people as a whole. In some cases a country's people also has to underwrite the construction of stupendously expensive modern pulp plants. Yet while these costs have to be met by all citizens, very few reap the profits. For example, in Chile — one of the most 'successful' cases of large-scale plantation development — ten years of government subsidies contributed to a state of affairs in which, in 1985,

> just three Chilean corporations held 70 per cent of the planting grants, plantation areas, and timber exports — a very uneven distribution of the costs and benefits of plantations (CODEFF 1991, cited in Sargent and Bass 1992).

Concentration of wealth implies concentration of power and dispossession of local communities. In Thailand, for example, industrial plantations have been 'an exceptionally efficient device allowing interests responsive to the world economy to annex supposedly "marginal" areas, smash the remaining local-oriented noneconomic or semi-economic pattern of livelihood and nature conservation there, and convert the fragments into 'resources' for global exchange. As land is concentrated and transformed into a substrate for eucalyptus, local villagers are cut loose to seek niches as producers, consumers, recyclers or commodities in the world economy' (Lohmann 1991).

A further problem is the risk of national dependence on a commodity prone to wild tumbles in price (see chapter 2). Indiscriminate planting of pulpwood trees or any other crop can lead to a glut of raw materials which, however beneficial it may be for paper manufacturers and users, makes their cultivation progressively less profitable. Indeed, tree crops on the whole are already chronically unprofitable in strict market terms, as otherwise they would not need so many state subsidies. But new risks are being added by the planting of millions more hectares of tree crops around the world in the next few years, which may put pulpwood into the bracket of other Southern primary commodities whose prices have fallen to persistently uneconomic levels (Instituto del Tercer Mundo 1989). Yet Southern countries committed to pulpwood exports, as to other commodity exports, are likely to have to continue exporting at ever-lower prices, competing among themselves for industrialized-country buyers. Indeed, the situation is even more serious for pulpwood than for annual crops, since it is not only much more expensive to cut trees prematurely than to plough a crop which has not yet matured, but also more difficult to return land to agriculture after trees — particularly eucalyptus — have been planted on it. In addition, the tree plantations in question may have been occupying the land for a number of years, raising financial losses even higher. A landscape of 'tree cemeteries' — masses of uncropped industrial plantations like those described by Pérez Arrarte (1995) — may be a real possibility in certain locations in the future.

Similar risks, of course, afflict pulp — the commodity which appears to be replacing pulpwood. Here price drops are likely to be especially serious for huge exporters such as Brazil and Indonesia, who have had to invest the same immense amounts in mill equipment as their Northern competitors, but who hold less capital to cushion potential losses when the market turns. Pulp production, like the large-scale industrial plantations with which it is associated, is likely neither to bring profits to the majority of a country's people nor to decrease vulnerability to economic domination by the industrial North.

Conclusions

Large-scale industrial tree plantations undoubtedly help the international pulp and paper industry secure stable supplies of raw materials. They are also capable of periodically making sizable profits for the huge conglomerates which plant them. They are not designed, however, to benefit Southern countries as a whole, their people or their environments. Although they normally destroy more employment than they create (see chapters 7 and 8), they nevertheless rely on subsidies extracted from large numbers of people to generate their profits. They do not help preserve land, forests, grasslands, or water sources, but rather exploit local natural advantages ruthlessly.

Neither Southern countries nor their local communities, therefore, should hope to benefit from the presence of huge plantation and pulp firms producing for export. On the contrary, they must be on their guard against the damage these corporations can wreak. While plantation tree roots may be within national territory, it is very unlikely that the roots of such companies will be.

Chapter 5

Actors behind the Scenes

The plantation boom this book describes is not the consequence of a conspiracy among a small group of like-minded industrialists, implemented with the careless ease of omnipotence. Nor is it the result of 'free-market mechanisms' or an inevitable and impersonal 'drive for economic development' which must now be made 'sustainable'. It is, rather, a complex social, cultural and political struggle featuring a multitude of agents with different interests, cultural inclinations, and motivations, acting against a constant and varied background of resistance. It is only by working in loose conjunction with each other (although usually in a contingent and sometimes uncoordinated and *ad hoc* fashion) that these actors can make paper and pulp machinery saleable, debt finance possible, political interests meshable, and centralization of resource control achievable. As a prelude to the case studies of Part Two, this chapter will introduce some of the most prominent of these actors, sketch something of their interactions, and outline a few of the strategies which shape their actions.

Pulp and paper firms

The most important customers for plantation fibre, of course, are paper manufacturers. Together, these firms form one of the top ten industrial sectors in the world, representing one per cent of the world's total economic output. The 65 leading producers of paper are listed on pages 106–113.

As chapters 2 and 3 have pointed out, the industry is dominated by the North, where as much as 90 per cent of the world's pulpwood, over four-fifths of the world's pulp, and between three-quarters and four-fifths of the world's paper is produced. It is dominated particularly by the United States, which boasts between a third and a half of the world's pulpwood production and a rate of paper production about equal to that of the next four largest-producing countries combined (Fernandez Carro and Wilson 1992, Wilson 1991, IIED 1995, *PPI* 7.1994, FAO 1995, van Hook 1994). Nevertheless, as chapter 3 argues, more and more pulpwood plantations are likely to be established in the South, with a good deal of manufacturing capacity following suit, and several large Southern conglomerates are already climbing into the world league tables.

Given the size of large paper firms — the sales figures of International Paper alone rank above the Gross Domestic Products of more than 75 countries — it is hardly surprising that many of the largest paper firms are important political as well as economic actors. Some firms, in addition, can take advantage of the political clout that comes with being a member of even larger conglomerates. Arjo Wiggins Appleton, for example, whose sales figures alone exceed the Gross Domestic Product of Honduras, is a subsidiary of the much larger British-American Tobacco firm, as is Aracruz's parent firm Souza Cruz (WRI 1994, *PPI* 9.94). Carter Holt Harvey, a firm with pine plantations in Chile and New Zealand, is run not only by International Paper, but also partly by Brierly Investments of New Zealand, a firm with US$4.75 billion in assets worldwide, mainly in the UK.

As the list of firms on pages 106–113 shows, many prominent paper producers supply themselves with the pulp, chips, and logs they need from timberlands that they own or lease. Others buy pulp on the open market from producers elsewhere on the list, or from other market pulp specialists such as the US's Alabama River and Parsons & Whittemore; Brazil's Bahia Sul Celulose; Sweden's ASSI Karlsborg and Iggesund Paperboard; Swaziland's Usutu Pulp; Spain's Sarriopapel y Celulosa; Indonesia's Wirya Karya Sakti; and Canada's Canfor, Cariboo, Crestbrook, Irving, and Malette (the first three of which involve joint ventures with Japanese paper interests) (IIED 1995, *PPI* 3.94). While Brazil is the South's leader in the pulp trade, Indonesia is currently the fastest-growing market pulp producer, with growth in the sector averaging nearly 29 per cent yearly between 1980 and 1991.

Still other corporations do not produce pulp and paper themselves, yet are involved in the wood trade or in the development of pulpwood plantations. Shell International, for example, began forestry operations in Brazil in 1980 and by 1991 was involved in plantation projects in Congo, Chile, New Zealand and South Africa, as well as research operations in Australia, Thailand, France, the US, the UK, and New Zealand (Bissio 1991). While Shell's long-term interest is partly in developing petroleum replacements from biomass (an interest shared by FLORAM in Brazil), its wood-fibre operations are currently linked to the paper economy. Itochu, similarly, has been involved in the wood chip trade in Southeast Asia and elsewhere. Traders such as Sweden's CellMark — which is to join Marubeni in marketing pulp produced by the giant new Tanjung Enim Lestari mill in Indonesia — meanwhile sell pulp and paper which other firms produce.

Consultancy companies

Playing a crucial indirect role in pulp and paper manufacture are a handful of Northern forestry and engineering consultancy firms whose business is promoting, investigating, planning, designing and setting up pulp and paper mills or logging and plantation operations.

Such companies include Canada's H. A. Simons, Reid, Collins and Associates, Sandwell, and SNC Lavalin; the US's Brown and Root, Babcock and Wilcox, CH2MHill, and Rust Engineering; the Nordic countries' Silvestria, Swedforest and ENSO/Indufor; Switzerland's and Britain's SGS Silviconsult; New Zealand's FORENCO Consultants; Australia's FORTECH; Germany's DFS Deutsche Forestservice; and France's Cirad and Chleq Froté.

Preeminent on the international scene is Finland's Jaakko Pöyry — the largest forestry and engineering consulting company in the world, with an estimated 40 per cent of the forest industry consultancy market worldwide and a turnover of more than US$300 million in 1994 alone. Pöyry, which has recently absorbed the large Swedish consulting firm Interforest, has over 60 offices in 25 countries around the world — 11 in Brazil alone — and thousands of employees, and has been involved in hundreds of major commercial forestry and pulp and paper projects in the last two decades across the Americas, Africa, Asia, Oceania and Europe (*FT* 8.3.95; Jaakko Pöyry 1994, n.d. a, b, c, d).

Firms such as Pöyry are typically in the vanguard of industrial tree plantation expansion. Wherever there are possibilities for industry growth, consultants are likely to be on the scene early, lobbying governments, evaluating forest and land resources, lining up contracts from close colleagues in 'aid' agencies, subcontracting lucrative work out to potential local allies, doing feasibility studies or market surveys, establishing tree nurseries, and designing or engineering factories. Relying on contracts both from state and international agencies and from the private sector, Pöyry and its fellow consulting firms serve as crucial go-betweens linking the interests of international and national business and officialdom and bringing together Northern machinery and techniques with Southern land and forests.

The more successful such consultancies are in setting up or expanding industrial forestry or pulp and paper sectors, the more work they are ensured in the future. Essential here are influential contacts both inside and outside government and international agencies. In 1994, for example, Pöyry, which had no previous experience in India, was selected over 15 Indian bidders to carry out World Bank forestry projects in Kerala and Uttar Pradesh. Surprise at this remarkable coup was somewhat lessened by the revelation that the person in charge of Bank forestry programmes in India was a former vice-president of the Jaakko Pöyry Group, Christian Keil. India's Inspector General of Forests, A. K. Mukerji, meanwhile, who had recently been a guest of Jaakko Pöyry in Finland, was reportedly preparing to open a branch of the firm in India upon his retirement from the civil service (*Nation* 27.11.94; *Statesman* 16.9.94). As Part Two will document, such coincidences are common. There are, moreover, virtually no official or professional sanctions which can be applied in the home countries of such consulting firms against their questionable practices abroad.

The population density of Northern consultants in the South grows when experienced consulting staff set up their own firms in Southern countries in which they have worked — thus giving themselves, and the Northern consultancies with which they are still linked, an inside track on future contracts. A US$1.5 million Asian Development Bank contract for developing fast-growing tree plantations in three Lao provinces bordering Thailand, for example, is being shared by Jaakko Pöyry AB of Sweden and the Vientiane-based Burapha Development Consultants, whose staff is partly Swedish (*DT* 9.94).

Providing additional support to pulp and paper development are consultancies which specialize in market forecasting, technical research and training. The UK's Pira International, for example, performs information and research services for the paper, packaging and publishing industries, while firms such as Wood Resources International and Jay Gruenfeld Associates of the US and Hawkins Wright Ltd. of the UK monitor pulp or wood markets. Organizations such as London's *Financial Times* also provide valuable research and information for the industry, as well as hosting inter-industry meetings.

Technology suppliers

Working together with forestry and engineering consultants for mutual benefit are the major Northern suppliers of pulp- and paper-making technology. Because these firms depend on sales of large-scale, expensive machinery, it is to their advantage too to support and lobby for expansion of large-scale monoculture tree plantations. The profits to be made are enormous, with the machinery costs of each new pulp mill running into the hundreds of millions of dollars.

Unsurprisingly, technology suppliers tend to be based in the same countries as forest industry consulting firms. Like the more prominent consultants, the larger technology suppliers maintain offices in Southern locations from Singapore to São Paulo, and often team up in multinational consortia to develop or market machinery.

Among the main world suppliers of pulping and bleaching equipment are Finland's Ahlstrom and Valmet-Tampella, Sweden's Kværner Pulping and Sunds Defibrator, the US's Beloit, Switzerland's Sulzer, and Britain's Black Clawson. Other major suppliers include:

- *Papermaking equipment*: Valmet (Finland); Beloit (US); Voith Sulzer Papiertechnik (Germany/Switzerland); Mitsubishi Heavy Industries (Japan) and Sunds Defibrator (Sweden).

- *Power and steam equipment*: ABB and Götaverken (Sweden); Ahlstrom, Tampella, Outokumpu and Valmet (Finland); Babcock & Wilcox (Canada); General Electric (US) and Voith (Germany).

- *Finishing and converting equipment*: Beloit (US); Sulzer (Switzerland) and Valmet (Finland).

- *Instrumentation, process control and automation*: ABB (Sweden); Allen-Bradley, Modicon, Measurex, Rosemount and Texas Instruments (US); Black Clawson (UK); Fischer & Porter and Honeywell (US); Kytola and Valmet (Finland); Siemens and Voith (Germany); Sulzer (Switzerland); and Yokogawa (Japan).

- *Pulp- and paper-making chemicals*: BASF, Dow and Texaco (US); Ciba-Geigy and Sandoz (Switzerland); Eka Nobel (Sweden); Kemira (Finland); Nalco (Austria); Grace Dearborn and SCM (UK); and Rhone Poulenc (France).

- *Chemical recovery equipment*: Ahlstrom and Tampella (Finland); Babcock & Wilcox (US); and Gotaverken (Sweden).

- *Woodyard equipment*: Kone and Sunds Defibrator Wood-handling (Finland); Iggesund and VME (Sweden); and Maschine-fabrik Andritz (Austria).

- *Environmental control equipment*: Babcock & Wilcox (US) and Voith (Germany).

Industry associations and alliances

Like their counterparts in other sectors, pulp and paper industrialists have long been aware of the need to join forces to achieve common goals. In addition to constant flows of personnel and day-to-day efforts at coordination among suppliers, buyers, consultants, machinery manufacturers, and so on, the industry holds regular conferences to consider general market prospects, technological developments, challenges from environmentalists and environmental legislation, and ways of capturing subsidies.

Such meetings — underwritten by industry groups and organized by such bodies as the *Financial Times* or paper industry journals — are often heralded by the obligatory disclaimer that they will, for example, 'comply with the antitrust laws applicable in the United States and the European Union'. This does not stop their participants from urging more inter-industry collaboration to deal with difficult 'issues that go beyond the immediate realities of our competitive marketplace'. Industry unity is vital, many of its leaders insist, to prevent consumption from peaking. They add that different companies should also pool 'our substantial R & D funds and expertise' to come up with new, more efficient technologies which can comply with environmental standards; must contribute more to environmental standard-setting and certification; and must learn to present a single public-relations front on environmental matters in order to build the right 'public perceptions'. A united front with customers in publishing is also crucial, many executives insist, and industry must take care not to let environmentalists push firms into destructive competition over such issues as recycling and chlorine-free paper production (Oberlander 1994; *Paper Europe* 1995).

In fact, of course, many industry alliances already exist. Associations such as the the Indonesian Pulp and Paper Association, Confederation of European Paper Industries (CEPI), the American Forest and Paper Association, the Council of Forest Industries of British Columbia, the Japan Paper Association, the Associação Brasileira de Exportadores de Celulose, the Thai Pulp and Paper Industries Association, and the Paper Federation of Great Britain have long represented corporate interests to governments and the public. Firms such as Beloit and Mitsubishi Heavy Industries have combined research efforts to come up with new papermaking technology. Northern manufacturers have also banded together to work out common strategies for marketing machinery in the South with the help of development 'aid' (*WW* 6.92).

Industry associations play a significant role in international politics. Sweden's pulp and paper associations, eager to gain more political clout in Brussels at a time when the industry was rapidly internationalizing throughout Europe — some 40 per cent of Swedish-owned paper production capacity is already sited abroad — were influential in persuading the country to join the European Union. The increasingly aggressive Canadian Pulp and Paper Association, meanwhile, is seeking closer alliances with paper users such as the *New York Times, Wall Street Journal*, and Knight Ridder Newspapers, and has set up an office in

Brussels to coordinate campaigns to secure European clients in the face of environmental protests against forestry practices in North America (*PRW* 1.95). The Finnish industry's jointly-owned Paperinfo, an 'information service company', monitors international markets and environmental groups, while the Finnish Forest Industry Federation mounts international public-relations blitzes. In 1990, an opinion-management organization called the Pulp and Paper Information Centre was set up in London by several inter-industry groups. The work of such alliances will be examined in more detail in chapter 6.

Bilateral agencies

Many of the Northern companies discussed in this chapter — whether forestry and engineering consulting firms or paper, pulp, wood or machinery companies — owe their survival largely to handouts from their governments for their work in the South. Without grants, cheap finance, and other subsidies, billion-dollar mills and tropical plantations — which can cost US$1,000 per hectare to set up (Pandey 1992) — would never make it into production.

Some of these subsidies, as will be discussed below, are transferred to multilateral agencies before being shunted to the business sector. In other cases, funds collected from Northern taxpayers are laundered through governmental bilateral 'aid' agencies before being channelled into the coffers of private firms. To take just one example, roughly half of Norway's 1994 aid budget was spent on Norwegian firms' goods and services (*DT* 20.10.1995).

Like the states to which they belong, bilateral agencies are nonetheless far from being mere creatures of their countries' corporations. The staffs of such agencies need to be responsive not only to pressures to subsidize national industries but also to imperatives to ensure their own institutions' survival, meet their governments' foreign policy objectives, and live up to their name by making at least some token gestures toward 'helping' the poor abroad.

Tensions within bilateral 'aid' agencies add a further dimension to their relationships with large corporations. In the Nordic countries, for instance, the 'aid' agency/consultancy/manufacturing firm relationship has often been mediated by what are known in Swedish as *närkonsult* or near-consultants. These senior figures, with wide experience not only in the forestry industry but also in government bureaucracies and multilateral and bilateral agencies, have normally neither worked for companies nor implemented projects. Rather, they have written terms of reference for agency projects. The combination of abstraction and narrowness which characterizes such documents has often provoked resentment even among the corporate consultants who have won the bids to carry them out. As one consultant remarked recently, 'even if we want to do something good or progressive, we are limited by how [the *närkonsult*] have written up the terms of reference . . . We think they're arrogant and they think we're stupid' (Usher 1994).

A number of bilateral agencies have played an important role in promoting monoculture tree plantations:

- Finland's FINNIDA has bankrolled plantation and pulp and paper mill planning; pulp mill equipment exports; feasibility studies; plantation trials; management and technical services and appraisals; preparation of projects for bank financing; and national forestry planning (including pulp plantation programmes) for the Philippines, Thailand, Sri Lanka, Tanzania, Nepal, Zambia, Kenya and other countries — all carried out by Finnish and Swedish firms (Jaakko Pöyry n.d. b, c, d).

- Sweden's SIDA has funded plantation planning; data collection and implementation; training; extension; national forestry planning; and appraisal — again, all carried out by Swedish and Finnish firms. Examples include the planning and implementation of fibre plantations for a 50,000 tonne-per-year pulp and paper mill in northern Viet Nam and inventory and exploitation plans for pulpwood plantations in Mozambique and Tanzania (Interforest n.d. a, b, c).

- Canada's CIDA has supported national forestry planning in Zaire and Peru and H. A. Simons's consultancies with private firms abroad.

- Japan's JICA has provided handouts for plantation research, planning and trials in order 'to encourage industrial plantations' in Chile, Uruguay, Paraguay, Indonesia, Thailand, Malaysia and other countries, while its Overseas Economic Cooperation Fund has subsidized Japanese corporate wood chip consortia (JICA n.d.)

- Britain's Overseas Development Administration has funded a 'social forestry' programme in India's Karnataka state which was used to convert commons and farmlands into industrial eucalyptus plantations, provoking heavy and widespread protests by those impoverished (SPS 1989).

Other bilateral agencies subsidizing exports which promote the pulp, paper and plantation industries include the US's USAID and Germany's GtZ.

Subsidies from such agencies are kept high partly through competition among Northern firms and countries to gain influence in particular countries, with each nation's agency striving to make its favoured corporations' services cheaper to Southern elites. 'Aid' deals lucrative to Southern elites — and involving subsidized work by Northern firms — are also often offered to Southern governments as bribes in exchange for buying weaponry or other goods from Northern countries (*Observer* 7.5.89, 6.2.94, 13.2.94; *Times* 6.3.94).

Proponents of the 'free market' and 'good governance' among the European and North American private and public sectors have so far been able to restrain themselves remarkably well from flinging accusations of 'socialism' or 'corruption' at such practices. As Walter Lippmann observed long ago, while the 'free market' is considered a useful tonic for the poor, the rich continue to reserve socialism for themselves. Meanwhile, Northern governments' practices of paying out backhanders and brokering backroom deals are advertised as respectably capitalist endeavours, providing, it is claimed, domestic jobs as well as increasing exports and foreign exchange. In such cases, the word 'corruption' is held by definition not to apply. Words such as 'reasons of state' or (at worst) 'sleaze' or 'conflict of interest' are pressed into service instead.

State investment and export credit agencies

Also helping to subsidize the international expansion of commercial pulpwood plantations with public funds are Northern government-supported investment corporations.

One such agency is Britain's Commonwealth Development Corporation (formerly the Colonial Development Corporation), or CDC. CDC invests taxpayers' money earmarked for 'foreign aid' in tree plantation or pulp and paper companies in Africa, Asia and the Pacific, especially in areas where private investors may be reluctant to tread. In recent years, CDC has drawn around 45 per cent of the over £150 million it invests annually directly from the British 'aid' programme; the remaining 55 per cent derives largely from profits made on seed money also taken from the 'aid' budget (CDC 1992).

Among the companies in which CDC invested in 1991 was Thailand's Soon Hua Seng Group, whose chief executive Kitti Damnoen-charnwanit had been arrested only one year previously for illegally logging a National Reserve Forest in preparation for planting eucalyptus (see chapter 13). Despite Soon Hua Seng's reputation for land scams, strongarm tactics, and environmental irresponsibility, CDC approved a loan of £15 million to the company specifically for eucalyptus plantings and a sawmill in an environmentally-sensitive catchment area in the east Thailand hinterlands in which land speculation and intimidation and eviction of smallholders was rife. CDC has also provided debt finance for a new papermaking plant Soon Hua Seng is planning for east Thailand. It has invested, in addition, in plantation or pulp and paper companies in Côte d'Ivoire, the Solomon Islands, Fiji, and Swaziland. In the latter country CDC is both the sole owner of the Shiselweni Forestry Company — which controls 14,000 hectares of eucalyptus and pine plantation — and, with South Africa's Sappi firm and the Swazi government, the joint owner of the Usutu Pulp Company, which produces solely for export. Although CDC is sponsored by ODA, Britain's official bilateral aid agency, and in principle is accountable to Parliament, in practice it is exempt from public oversight. Its officials contemptuously reject outside requests for detailed information on its activities, whether they come from the British taxpayers whose earnings they use or nationals of the countries affected by its investments (CDC 1992, Adlard 1993, Chalker 1992, Magee 1994).

Other Northern government-funded agencies and initiatives also help finance the sale of goods and services to pulp, paper or plantation companies in the South and elsewhere. Finland's Premixed Concessional Credit Scheme, for example, helped underwrite an increase in the export of Finnish machinery to Indonesia and Thailand from US$21 million in 1990 to $211 million in 1993 at a time of lagging domestic sales; most of this money went for forestry and pulp and paper equipment. The Canadian government and the US Export-Import bank, meanwhile, have offered millions of dollars in concessional loans to firms such as Thailand's Panjapol pulp and paper group. Sweden's Board for Investment and Technical Support (BITS), Canada's CIDA Inc., and the Japanese Export-Import Bank are also active in subsidizing their countries' exports to Southern pulp and paper firms. The US's Overseas Private Investment Corporation (OPIC), similarly, has financed American logging operations in Siberia's Far East. Together with the US Import-Export Bank, OPIC is also helping to lubricate an intergovernmental deal which will result in US industry's sending billions of dollars' worth of pulp and paper, logging and other machinery to Siberia in exchange for Russian wood (Finland National Board of Customs 1990–3, Olsson 1995).

Multilateral agencies

Still more subsidies flow directly or indirectly to the pulp, paper and plantation industry, including Northern consultancy, construction and machinery firms, from multilateral development banks (MDBs) such as the Asian Development Bank (ADB), African Development Bank, Inter-American Development Bank (IDB), the World Bank and the European Bank for Reconstruction and Development. Of the US$40 billion which MDBs loan annually on favourable terms, a considerable mite goes toward planning and reviewing pulp or plantation operations and facilitating investment in forest industries.

Between 1984 and 1994 alone, for example, the World Bank provided cheap finance for the establishment of 2.9 million hectares of tree plantations, largely for commercial purposes, at a cost of US$1.416 billion. Countries affected have included China, India, Bangladesh, Algeria, Tunisia, Zimbabwe, Kenya, Peru and Poland (World Bank 1994b). Multilateral development banks have provided support for analyses describing how to build up Indonesia's pulp and paper industry (1984 and 1987–8, involving the World Bank and ADB), studies of investment opportunities in Latin America, Viet Nam and Nepal (1981–2, 1990–1, and 1986–present, involving IDB and ADB), and forestry development plans for Cameroon, Central African Republic, Congo, Ethiopia, Ghana, Nigeria, Sudan, Bangladesh, Bhutan, Laos, Nepal, Pakistan, Papua New Guinea, the Philippines and Sri Lanka (Jaakko Pöyry n.d. b, c, d). Current World Bank schemes include a Forest Resource Development and Protection Project under which US$200 million is being loaned to China for, among other things, intensively-managed plantations of larch, pine, eucalyptus and other species which would support a more capital-intensive, centralized, wood-based paper industry (World Bank 1994c).

Although World Bank loans for pulpwood plantations are often concealed within larger loans for forestry or agroindustry or disguised under the euphemism 'resource expansion', blatant handouts to industry are common. In Minas Gerais, Brazil, for example, a Bank-financed project is inducing local farmers to raise eucalyptus saplings to provide raw materials for the pulp and paper industry and fuel for steelworks, thus diverting and degrading soil resources which would otherwise be used for food production (*WRR* 2.94). And in Kenya, the Bank loaned US$19.9 million for a forestry development project promoting tree farming at a time when the International Finance Corporation (IFC), the Bank's private investment arm, had invested $86 million in Kenyan pulp, paper and packaging production (including a TetraPak packaging plant).

MDBs, through their reassuring presence and groundbreaking operations, have also encouraged ordinary commercial banks to get involved in tree plantations or pulp and paper manufacture in the South — as lenders, guarantors of export credit agency loans, or advisers on the raising of debt finance. By financing export-oriented industrial development, moreover, they stimulate local markets for packaging and business papers.

Some of the money used by MDBs to subsidize the pulp and paper industry is derived from past and present contributions from Northern taxpayers. The rest is 'donated' by the South itself in the course of paying off debts incurred through the construction of previous MDB-initiated projects, many of which are now acknowledged by MDBs themselves to have been boondoggles (Wapenhans 1993, Rich 1994). The specific destinations of the money which cycles through MDBs vary. Some of it is skimmed off by Southern elites. Huge amounts, however, materialize almost immediately in the bank accounts of Northern consultants, contractors and suppliers.

The mechanisms by which this happens are not hidden. The World Bank, for example, has a policy of financing components of its projects that must be purchased with foreign exchange — on average, about 40 per cent. With about US$23 billion committed annually, some 30,000 contracts are awarded to firms annually by borrowers of World Bank funds, 70 per cent for goods and equipment, 20 per cent for civil works, and ten per cent for consultancies. This structure encourages close links between the Bank and representatives of private firms who are seeking outlets for their equipment or expertise.

Northern firms in particular enjoy a number of resources which help them build such links. The US Department of the Treasury, for example, has instructed its representatives on the boards of directors of MDBs 'to make [US] business development and exports one of their highest priorities within the MDBs' and to impress on the MDBs the virtues of 'one-stop shopping' at US firms. The UK's Department of Trade and Industry (DTI) employs no less than six desk officers to provide the British private sector with pointers on how to 'gain business' from multilateral development bank-funded projects through cultivating the right staff members, visiting World Bank headquarters, and so forth. In addition, a whole satellite industry of consultants — many of them former World Bank staff or the spouses of current staff — has sprung up in Washington, London, and other capitals to monitor MDB procurement and offer inside information and services to private firms; and MDBs and Northern governments hold regular meetings in Northern cities to help the Banks and prospective Northern contractors get to know each other (World Bank n.d., *Treasury News* 18.11.93, DTI 1994).

The result is predictable. According to Britain's Department of Trade and Industry, in the financial year ending in June 1994, companies in the US, UK, France, Germany and Japan captured over 46 per cent of the cash value of the contracts awarded by the World Bank (DTI 1994), or more than US$10 billion. Some 64 per cent of the value of IDB disbursements, meanwhile, went to firms in five wealthy Northern countries: US, Japan, Germany, France and Italy. As the UK's Overseas Development Administration (ODA) complacently observes, 'British firms and suppliers receive orders for goods and services that are used in aid projects far exceeding the value of Britain's contributions to [multilateral] agencies' (ODA 1992).

Much of the money captured by MDBs thus follows a roughly circular pipeline from Northern taxpayers to Northern governments to MDBs to Southern governments to Northern companies, leaking copiously through cracks along the way. Replenishments are meanwhile being continually pumped from both Southern and Northern catchment areas directly to MDBs to be periodically injected at the top of the cycle; the Southern contribution to these topups makes up a significant part of the current net annual US$50 billion capital drain from South to North (Rich 1994).

The damage inflicted on the Southern catchment area in the process is considerable. As a price for the privilege of being able to insert taps into the circular financial flows to which their countries already contribute so much, Southern elites are compelled to commit themselves to projects and austerity programmes of staggering economic, social and environmental destructiveness. For nearly fifty years, in addition, Southern elites have been prodded by MDBs to set up or augment state institutions which nurture the growth of local commercial elites — including local contractors who quickly learn to feed at the MDB trough themselves. One result has been the creation of a permanent infrastructure for negative redistribution of wealth both within Southern countries and between South and North. The North-South wealth ratio has increased from 10:1 in 1948 to 30:1 in 1960 and 60:1 in 1990 (Rich 1994, Payer 1991, Adams 1991, Oxfam 1994).

Other multilateral agencies such as the Food and Agriculture Organization (FAO) and the United Nations Development Programme (UNDP) cooperate closely with MDBs and private sector consultants in using public funds to provide 'free lunches' for the plantation and pulp and paper elite. Such bodies are particularly helpful in studying business opportunities in the forestry sectors of Southern countries and planning rural development and forestry programmes which can feed the industry. FAO has also performed the valuable service of discrediting, through bogus science, the environmental soundness of certain indigenous land-use patterns, including various types of swidden agriculture, which were seen to be in competition with industrial forestry interests. As anthropologist Anders Baltzer Jørgensen (1979) notes, FAO 'took over a special branch of silvicultural opinion: the one formerly contained in the forest departments of earlier colonial governments' as part of a post-colonial project of saving the 'productivity of the tropics for the world as a whole, but specifically for the occidental world' (Spencer 1966).

Another intergovernmental agency, the Institute of Applied Systems Analysis based in Austria, meanwhile provides computer-assisted databases for industry about timber supply in Siberia, Europe and the rest of the world. The World-Bank-sponsored Center for International Forestry Research (CIFOR), a new branch of the Consultative Group on International Agricultural Research (CGIAR) located in Bogor, West Java, with 100,000 hectares of research forest in east Kalimantan, may also be an institution to watch. Experience suggests that there will be heavy pressures on CIFOR to devote much of its time to helping fit tree characteristics to industry needs through genetic research in much the same way that the International Rice Research Institute in the Philippines (another part of the CGIAR network) has helped to adjust rice to the chemical regimes of the Green Revolution (Jaakko Pöyry n.d. b, c, d; Interforest n.d. a, b, c; Moniaga 1993; IIASA 1994).

As the case studies of Part Two will propose, one of the most important functions of subsidies from multilateral agencies to Northern consultants for initial studies is that they often lead to knock-on contracts for the firms involved. After landing a contract in 1984 from the World Bank to make recommendations for the pulp and paper industry in Indonesia, for example, Jaakko Pöyry's consultancies for the Indonesian private sector mushroomed. Occasions such as the 1992 Earth Summit in Rio de Janeiro also provide rich opportunities for firms such as Pöyry to lobby for the diversion of public funds to their corporations in the guise of 'environmental aid'.

In summarizing the role of multilateral agencies in the promotion of pulp and plantations, it is hard to improve on the words of Jack Westoby, the late head of forestry at FAO. In 1975, Westoby described how international 'aid' had helped in 'identifying for foreign capital those forest resources suitable for exploitation' and in many cases had 'borne a substantial part of the cost of making inventories of those resources':

In not a few cases it has compiled the data, and helped provide the justification, for international financing agencies to provide loans to create some of the infrastructure needed to assist the penetration of foreign capital. It has helped to train some of the manpower to be placed at the service of foreign enterprises [and] assisted some irresponsible governments to alienate substantial parts of their forest resource endowment ... the growing interest in, and acceptance of, forestry projects [by development agencies] had little or nothing to do with the conversion of the development establishment to [foresters'] idea that forestry and forest industries had a significant and many-sided contribution to make to overall economic and social development. It had everything to do with the fact that many of the rich, industrialized countries needed, and needed badly, new wood material resources; and their forest industries, their equipment manufacturers, together with miscellaneous agents and operators, scented golden opportunities for profit in those underdeveloped countries with forest resources. This was the dominant consideration which determined the location, shape and direction of forest and forest industry development projects. .... The international financing agencies knew what foreign investors wanted, and the multilateral and bilateral agencies fell into line (Westoby 1987).

National governments

Some of the most important subsidies for the plantation, pulp and paper industries are furnished by national or provincial governments, often at the behest of international agencies.

State agencies set up originally under the aegis of the World Bank, for example, such as Thailand's Board of Investment, have provided tax writeoffs, technology import exemptions, and rent-free loans to industry, as well as infrastructure. Fiscal incentives including subsidies for plantation establishment, tax exemptions, and grants, have also been extended by governments to commercial plantation firms in Brazil (1966–1987), Argentina (from 1974), Chile (from 1960), Paraguay (1980–4) and Uruguay (1960–78) (Pandey 1992, Sawyer 1993).

State forestry departments in Asia, meanwhile, redistribute vast swathes of land over which they have jurisdiction from their occupants to industry, often charging miniscule rents or stumpage fees. In Indonesia, for example, where 70 per cent of the land area is managed by the state forestry bureaucracy, industry is charged as little as US$0.30 per hectare per year for the use of plantation land. In both Indonesia and Thailand, special replanting funds are shunted selectively to plantation businesses, and in the former, plantations are further subsidized by logging revenues recycled through the state (see chapter 13). The smaller the rents, stumpage fees and timber prices levied by forestry departments, of course, the more resources have to be diverted from other sectors into forest management. Among the losers are programmes which might promote non-pulp-plantation, community-based land uses.

Costs of land and labour are also kept down through subsidies provided to military and police forces by local and foreign taxpayers. The minimum legal daily wage in Yogyakarta in Indonesia, for example, is US$0.60 and in West Nusa Tenggara $0.72; plans by a Japanese investor who proposed to pay workers $2.40 were rejected by officials of the Investment Coordinating Board on the grounds it would damage Indonesia's cheap labour advantage (*DTE* [London] 6.1993). Where customary lands have to be seized forcibly by the state before being turned over to industry, and where smallholders and migrant labour are used essentially as slave labour, repression — much of it underwritten by foreign training programmes and foreign military hardware — is a particularly important ingredient in the economic mix. Export taxes on staples can also help subsidize cheap labour for foreign investors. State university forestry faculties — many of them staffed by foresters trained in industrial forestry in countries such as Finland, Canada and the UK — can meanwhile be relied upon to provide useful lobbying and technical support for commercial schemes.

Competition with Northern wood-producing nations' own subsidizing apparatuses, moreover, is tough, forcing Southern nations to dig even deeper into their pockets to attract investors. In 1988, for example, the Canadian province of Alberta presented the Japanese firm Daishowa with C$70 million worth of infrastructure to help induce it to build a $579 million bleached kraft pulp mill in the middle of a forest. The Japanese-controlled Al-Pac plant in Alberta, meanwhile, has received, in addition to $75 million worth of infrastructure and construction finance, a $275 million subordinated income debenture — not to be repaid unless and until the mill is profitable — and a $125 million standby debenture for future expansion. The timber subsidy for the plant is an additional $78.7 million yearly. In addition to making a down payment of $125, then, each citizen of Alberta is donating an additional $28 per year for the privilege of allowing Al-Pac to use up the province's resources (*TN* 5.94, WCWC 1994, Pratt and Urquhart 1994). A further $47 million has been committed by Canadian governments for public relations for overseas forest industries extracting Canadian pulpwood (MacIsaac and Champagne

1994). According to forest economist Randall O'Toole, the US government showered $7.2 billion in tax breaks on the US timber industry between 1980 and 1989 and fed it $449 million worth of below-cost timber in 1992 alone (Letto 1994). Forest and paper industry Political Action Committees meanwhile help fund US legislators' election campaigns. Many originally state-owned Nordic forest industries which have been recently partially or fully privatized have also benefited from decades of government-enforced privileges.

So widespread are such corporate welfare programmes, indeed, that even the government of a relatively poor US state such as West Virginia can be prevailed upon to offer $200 million in construction loans, $60 million in road improvements and further millions in tax incentives to help the world's largest builder of pulp and paper mills, British-based Parsons & Whittemore, turn several hundred thousand hectares of Appalachian hardwood forests into pulp. Still further subsidies were secured by West Virginia governor Gaston Caperton, who, after receiving campaign contributions from Parsons & Whittemore, pressured US government regulators to allow the firm to pollute the Ohio River without baseline studies of dioxin levels being done (NFN 1995). George F. Landegger, Parsons & Whittemore chairman, had summed up the thinking behind such subsidization during earlier dealings in Canada:

> Adam Smith's invisible hand will not build pulp mills, particularly in the kinds of forest areas that remain available in Canada for exploitation. It requires a great deal of coordination between private interests and public interest, and the public interest being willing to take a risk for development. That risk has primarily political dimensions for the public sector, but financial dimensions for the private sector (Pratt and Urquhart 1994).

Thai economist Pasuk Pongpaichit offers a less mystified view of the matter:

> Economic theory tells us it's all right to subsidize education because it benefits the whole society. But while eucalyptus and pulp and paper industries earn profits for some, they cause problems for society. Therefore, economic theory tells us, they should be taxed. But instead the government does the opposite. This is a matter of influence and power (Pasuk 1995).

Research institutes and NGOs

A final type of subsidy comes from governmental and non-governmental research institutes which provide free or low-cost ideological or scientific backing for the plantation boom. One example is Australia's state-supported Commonwealth Scientific and Industrial Research Organization (CSIRO), which brings together academics and industry to conduct research and development on hardwood plantation techniques. Another is the Swedish Pulp and Paper Research Institute, which is supported by both industry and government. Many state universities in the US and Europe — including the University of Idaho, Oregon State University, and the University of New Brunswick — meanwhile accept research support from the pulp and paper industry (Cromer and Kile 1993). Of course, foresters and scientists with no direct ties to industry also contribute heavily to the industry-friendly component of the 'two libraries' which chapter 4 has described.

On the non-governmental side, the UK's International Institute for Environment and Development has helped Shell plan and legitimize a plantation project in east Thailand, since cancelled due to local resistance, and has also conducted a favourable mid-term review of Jaakko Pöyry's plantation-promoting Forestry Sector Master Plan for Thailand. In the 1980s, the IIED's US counterpart, the influential World Resources Institute, trumpeted the Aracruz plantation operation in Brazil as a model for forestry development. More recently, the World Wide Fund for Nature has collaborated with Shell on an 11-volume review of, and guidelines for, tree plantations. On a more rarefied level, independent researchers such as Norman Myers have lent credibility to the plantation boom by sketching grandiose and highly theoretical schemes for carbon-fixing through 'grand-scale' tree-planting (Myers 1991). Think tanks such as Resources for the Future in Washington meanwhile provide strategic information of interest to the industry on the long-term availability of industrial wood. No less important are NGOs, foresters and economists who, although acquainted with and concerned about the problems of the pulpwood plantation boom, find it strategic to defer to industry on the issue (Sargent 1990a, Sargent et al. 1992, Shell/WWF 1993). The role of such experts and NGOs in supporting the industry will be probed at greater length in the next chapter.

Conclusion

This chapter has profiled some of the leading protagonists in the drama of pulp and paper. It has also suggested some of the mechanisms by which they coalesce, capture subsidies, and — against a background of constant challenge and resistance which will be described more fully in Part Two of this book — act in a way which results in the expansion of pulpwood plantations in the South.

These mechanisms both presuppose and facilitate an ability to shift resources within a regionally and globally organized system. In order to be able to make effective inroads into rural societies in (say) Kalimantan or Bahia Sul, pulp and paper interests must be able to tap revenues which government bureaucracies have collected from office workers or landowners in Helsinki or Quebec, or cite arguments which the World Bank, using debt service payments from the Philippines or Mexico, has hired hack consultants from New Delhi or Virginia to rehearse and polish. Yet applying such resources effectively across extensive geographical and social landscapes presupposes that, to a certain extent, these inroads have already been made, that a far-flung political infrastructure sensitive to the needs of pulp and paper interests is already in place. Moreover, it is only when such inroads have already been started, and local barriers have already been breached, that the pulp and paper industry can gain access to the fresh flows of subsidies which make its continued expansion, and that of the globalizing system it relies on, possible.

Here, as elsewhere, globalization and subsidization advance hand in hand; neither can succeed without the other. Trade has not only continued to follow the flag but has also continued to help keep it flying. 'Trade', however, covers a greater multitude of activities today than formerly, and the 'flag' may bear the logo of a development bank, foreign charity, agriculture ministry, local police force, investment promotion board, or any other of the plethora of contemporary official institutions which serve elite interests without the possibility of democratic scrutiny ever even having to be raised.

Part Two will spell out in more detail how resource transfers between expanding public and private sectors in Southeast Asia, Japan and the West subsidize the pulp and paper industry. In particular, it will discuss how, in Asia, commons are dismantled and the fragments reassembled into a capitalist structure; how, in Latin America, the emergence of locally-appropriate alternatives to industrial plantations is being thwarted; and how, nearly everywhere, land and water are being expropriated. The next chapter will preface these studies by examining how an increasingly global system has taken on another challenge: managing resistance to its expansion. In particular, it will examine how globalization both enables, and is enabled by, attempts to block potential alliances between grassroots groups fighting monoculture pulpwood plantations in the South and environmental and social activists elsewhere.

TABLE 5.1: The top 65

Company (home country) (includes 1995–6 mergers and buyouts)	1994 paper output (mt/y)	1994 sales (US$m)
1. International Paper (US)	8.5 (est.)	16,530 (est.)
2. UPM-Kymmene (Finland)	6.6 (est.)	10,638 (est.)
3. Stone Container (US)	6.5	5,749
4. Georgia-Pacific (US)	6.1	12,738
5. Stora (Sweden)	5.6	6,337
6. ENSO (Finland)	5.5	5,551
7. SCA (Sweden)	5.3	7,104
8. Nippon Paper (Japan)	4.8	9,678
9. Champion International (US)	4.6	5,318

world paper producers

Remarks

Owns or controls over 28,600 sq km of timberlands for pulp and paper. Manages lands, through subsidiaries and shareholdings, in New Zealand, South Africa, Chile and the Southern US. Exports logs and hardwood chips to Japan, China, etc. Operates in 26 countries, sells in over 130. Large producer of market pulp, board, and printing, writing, and kraft papers. As one of the US's top ten polluters, the firm has been nicknamed 'International Pollution' and is known for worker safety and health violations and an aggressive anti-union stance. Major buyer of pulp. Recently took over Carter Holt Harvey, which holds 3,200 sq km of pine plantations in New Zealand and nearly equivalent pine holdings in Chile, and has also absorbed Federal Paper Board, a big market pulp maker.

Owns nearly 10,000 sq km; imports wood fibre from Uruguay and has planned Turkish plantations. Producer of magazine and fine papers and newsprint.

Makes board in 17 countries; has tried to log and plant pulpwood in Central America. A leading target of protest over its hostility to environmental initiatives.

No. 1 producer of communication papers, No. 2 producer of market pulp; big exporter. Owns or controls 26,600 sq km of timberlands. Founding donor of Center for Defense of Free Enterprise, leading think tank for the anti-environmentalist Wise Use Movement. Involved in joint paper venture in Indonesia, has interests in Brazil. Growing target of US protest.

Controls 15,000 sq km of land in Sweden, 6,000 in Canada, as well as around 2,000 sq km of eucalyptus plantations in Portugal and 145 sq km of pine in Chile. Produces in nine countries. Canadian operations have been criticized for herbicide use and clearcutting.

Partially state-owned. Top European packaging producer. Owns 4,750 sq km of Finnish land; also buys wood from private owners. Involved in Kalimantan plantations and Java packaging industry. Uses old growth from Finland, Russia.

Owns 23,000 sq km of forest land (much of it planted to exotic monocultures) in Sweden and also buys from private owners. Imports from Russia and has operations in Austria, France, Germany, Italy, and the Netherlands.

Has mills, plantations, or both in Brazil, Canada, Chile, Indonesia, New Zealand, Portugal, and the US. Formed from Jujo/Sanyo-Kokusaku merger.

Owns or controls 24,900 sq km of timberlands; is involved in plantations in Indonesia and Brazil, where it is the 4th largest papermaker and 6th largest pulp maker. Owns 400 sq km of eucalyptus in Brazil, where it is reportedly planning a 1000-sq-km export wood chip plantation. Target of recent protests in US.

Company (country) (includes recent mergers)	1994 output (mt/year)	1994 sales (US$m)
10. Kimberly-Clark (US)	4.6	12,114
11. James River (US)	4.1	5,400
12. Weyerhaeuser (US)	3.9	10,398
13. Jefferson Smurfit (US)	3.7 (est.)	3,750 (est.)
14. New Oji (Japan)	3.3	7,629
15. Union Camp (US)	3.3	3,396
16. Jefferson Smurfit (Ireland)	3.2	2,562
17. KNP BT (Netherlands)	3.2	7,223
18. Sappi (South Africa)	3.0	2,199
19. Fletcher Challenge (NZ)	3.0	4,818
20. Temple-Inland (US)	2.8	2,938
21. MoDo (Sweden)	2.6	2,625
22. Mead (US)	2.6	4,558
23. Daishowa (Japan)	2.5	3,077

Remarks

Has recently absorbed Scott Paper, which was forced to drop Indonesian plantation scheme in 1989 and has been attempting to sell its lands. Operates in Thailand, Honduras, Mexico, Costa Rica, China, South Africa and nearly 20 other nations. Owns or controls more than 36,250 sq km of timberlands, including plantations in Spain. Controls half of world market for tissue. Large buyer of market pulp, including eucalyptus pulp.

Has operations in 13 countries; owns 15,200 sq km of timberlands; has a five per cent share in Aracruz and an 86 per cent share in Jamont. Top napkin producer.

World's largest market pulp producer. Owns 22,400 sq km of forest land in the US; leases or controls an additional 75,000, mostly in Canada. Half of its timberlands are intensively-managed plantations. Exports to Asia, sells timber from Indonesia, and has interests in Siberia. Target of major recent protests in US.

Makes newsprint, board; operates in France, Mexico, Austria; moving into China.

Tied to Mitsui trading house. Helps operate hundreds of sq km of plantations in Australia, New Zealand, Fiji, Papua New Guinea, Viet Nam, Brazil, Thailand and Chile. Imported chips from the ChipDeco scheme in Kalimantan; runs mills in Canada and New Zealand. Formed from merger of Oji and Kanzaki Paper.

Manages intensive plantations in southern US. Operates in seven nations.

Schemes in 13 countries, including Colombian eucalyptus plantations.

Operates in nine countries.

Exports both market pulp and paper. Runs mills using South African pulp in the UK; has bought large paper firms in Germany and the US; is looking to manufacture in the Far East. Owns 4,000 sq km of plantations in South Africa and recently contemplated a joint plantation venture in Mozambique.

Owns or has cutting rights on 33,862 sq km in the Americas, Oceania. May expand in Chile, Brazil, Malaysia, UK and invest in China with Indonesian partner.

Operates in two countries; owns or controls 7,476 sq km of timberlands.

Controls 10,300 sq km of Swedish forest land and the holdings of the German firm PWA. Imports wood from Russia, pulp from Alberta. Large market pulp producer.

Operates in Argentina and Chile; owns or controls 5,412 sq km of timberlands.

Tied to Marubeni group. Imports 60% of its raw fibre from overseas, including Australia, Canada, Chile, Russia, Thailand, and the US. Imports pulp from Brazil, Canada, Chile, Finland, Portugal, Sweden, and the US. Has invested in mills in Canada, Thailand, the US. Aggressive foreign investor; big target of environmental protest.

Company	1993 output (mt/year)	1993 sales (US$m)
24. Westvaco (US)	2.5	2,608
25. Boise Cascade (US)	2.5	4,140
26. Abitibi-Price (Canada)	2.4	1,546
27. Bowater (US)	2.1	1,359
28. Arjo Wiggins Appleton (UK)	1.9	4,465
29. Willamette Industries (US)	1.9	3,008
30. Haindl (Germany)	1.8	1,404
31. Riverwood International (US)	1.7	1,283
32. Daio Paper (Japan)	1.6	2,623
33. Cartiere Burgo (Italy)	1.5	1,444
34. MacMillan Bloedel (Canada)	1.5	2,892
35. Gaylord Container (US)	1.4	784
36. Avenor (Canada)	1.4	1,389
37. Norske Skogindustrier (Norway)	1.4	1,299
38. Honshu Paper (Japan)	1.3	4,600
39. Metsä-Serla (Finland)	1.3	1,814
40. Domtar (Canada)	1.3	1,568
41. Hansol Paper (South Korea)	1.2	1,307

Remarks

Paperboard and packaging maker; holds a Brazilian subsidiary (Rigesa). Exports to 69 countries. Ranked first in US forest products industry in release of toxic chemicals in 1988–89; exploits Appalachian, Mississippi valley forests.

Owns or controls 24,000 sq km of timberlands; founding funder of Center for Defense of Free Enterprise. Recent target of protests in Maine and elsewhere.

Specializes in newsprint; controls or owns 95,772 sq km of timberlands.

Makes newsprint, kraft, and printing and writing papers in two countries.

Controls eucalyptus plantations in Iberia.

Manufactures only in the US. Large market pulp producer.

Manufactures in three countries.

Makes folding food and drink cartons; manufactures in 13 countries.

Holds chipping firm in US; supplied also from Brazil, Thailand, and 650 sq km of Chilean plantations.

Relies heavily on pulp imports.

Manages 15,000 sq km of timberlands. Plantations and pulp mills in SE US and subsidiaries in Australia, UK, Japan, The Netherlands. Vancouver Island logging has provoked bitter conflict. Makes containerboard, packaging, newsprint, pulp.

Manufactures paperboard and packaging only in the US.

Has access to 114,000 sq km of forests; favours clearcutting. Exports pulp to Japan, China and Europe. Was forced to close large pulp plant on Vancouver Island in1993 due to low prices. Big recycled newsprint maker.

Buys most wood on open market but also owns 2,140 sq km of own land.

Pioneered chipping of natural forests of Papua New Guinea, where it now holds 100 sq km of plantations. Involved in Brazil, Chile, Thailand, Canada, Hong Kong. Member of Oji industrial group and Daiichi Kangyo financial group.

Operates in seven countries; imports old-growth timber from Russia.

Operates in Canada only; controls or owns 60,000 sq km of timberlands.

Manufactures only in Korea but uses imported fibre.

Company	1994 output (mt/year)	1994 sales (US$m)
42. Amcor (Australia)	1.2	4,266
43. Sonoco Products (US)	1.2	2,541
44. Cascades (Canada)	1.2	1,263
45. INAPA (Portugal)	1.2	216
46. Settsu (Japan)	1.1	1,195
47. AssiDomän (Sweden)	1.1	2,172
48. Mayr-Melnhof (Austria)	1.1	942
49. Yuen Foong Yu (Taiwan)	1.0	634
50. David S. Smith (UK)	1.0	1,576
51. Consolidated Papers (US)	1.0	1,028
52. Klabin (Brazil)	0.9	1,005
53. Potlatch (US)	0.9	1,470
54. Mondi (South Africa)	0.9	875
55. Repap (Canada)	0.9	1,138
56. Hokuetsu Paper Mills (Japan)	0.8	1,084
57. Mitsubishi Paper (Japan)	0.8	2,042
58. QUNO (Canada)	0.8	411
59. Cheng Loong (Taiwan)	0.8	474
60. Rengo (Japan)	0.8	2,749
61. Chuetsu Pulp & Paper (Japan)	0.7	877
62. Shin Ho (South Korea)	0.7	1,220
63. Indah Kiat (Indonesia)	0.7	483
64. Noranda Forest (Canada)	0.6	1,319
65. Holtzmann Papier (Germany)	0.6	443

Remarks

Integrated packaging/paper firm operating in China, Indonesia, Malaysia, Hong Kong, Singapore, South Korea, nine other nations. Target of Australian protests.

Operates in 250 locations in 24 countries.

Operates in four countries.

Takes advantage of government-promoted plantations.

Tied to Sumitomo Group. Has interests in Brazil, Chile and Thailand as well as mills in Portugal, Spain, and the southeastern US.

Large forest owner and kraft paper producer operating in ten countries.

Board specialist operating in eight countries.

Holds shares in, and trades machinery to, Indonesian and Thai firms.

Holding company operating in four countries.

Operates only in the US but exports to Mexico, Asia and South America.

Latin America's top integrated forest products firm; owns 3,333 sq km; 39 per cent of sales are abroad.

Makes market pulp, board, and printing and writing papers only in US.

Plans to double pulp capacity after 2000; interests in UK and Portugal.

Owns 65,550 sq km of timberlands, an area the size of Sri Lanka.

Tied to New Oji, which holds shares; interests in Southeast Asia.

Holds 100 sq km of Chilean plantations; interests in Southeast Asia and Canada; art and speciality papers a focus.

Operates only in Canada.

Interests in Southeast Asia; operates in two countries; imports fibre.

Makes board in Indonesia, Thailand, China, Singapore, and Malaysia.

Associated with Oji group; makes newsprint and fine and printing papers.

Operates in three countries; involved in Thai newsprint mill.

Runs pulp mills, plantations and native forest concessions in Sumatra.

Controls or owns 57,750 sq km of timberlands, operates in two countries.

Operates only in Germany.

Sources: Olsson 1995, Lamb 1992, JATAN 1994, Marchak 1992, corporate reports, Rahikainen et al. 1995, PPI 9.95, FT and other newspaper reports.

Chapter 6

Managing Resistance

If the pulp and paper industry needs subsidies in order to be able to expand, it also needs to be able to manage opposition. Negative publicity about deforestation, dioxin, landfills, and so forth — to say nothing of confrontations with local people over the use of land and water — can threaten investment, demand, growth, market share and profits alike. Finding ways of taking the offensive on social and environmental issues is thus a preoccupation of nearly everyone in the industry.

Here again, global reach can help. Just as the industry uses world-spanning infrastructure, bureaucracy, and cultural influence to find sources and conduits for the subsidies it attracts, so it acts across wide geographical and cultural distances both to bring a wide variety of resources to bear on opponents and to try to divide them from each other more effectively.

Non-threatening resistance

From industrialists' point of view, some opposition to plantations or pulp and paper mills, while annoying and inconvenient, is relatively easy to get rid of or get around. It does not threaten their most fundamental interests either because it is scattered and localized or because the challenges it poses can be eliminated without overwhelming sacrifices, simply by redistributing resources from one part of a corporation to another. In such cases the industry tends to adopt one or more of the following strategies:

- *Wage economic or cultural war on pockets of resisters until they give up the struggle.* Smallholders who do not acquiesce in plantation or contract farming schemes may find themselves surrounded with fast-growing trees, their access to roads or rights-of-way cut off. Religious leaders or government employees may be induced to harrass plantation opponents or their families. Where political circumstances permit, resisters may also be isolated by being labeled 'Communists', 'traitors', or 'anti-development'. In areas coveted by speculators hoping to sell land to Shell (Thailand) for a plantation in the late 1980s, gambling schemes were even set up to relieve plantation opponents of their money in order to encourage them to sell out.

- *Buy off potential sceptics or resisters with money, land, goods, jobs or status obtained through the proceeds of operations elsewhere.* Bribes may be passed out, sports grounds or clinics built, official titles bestowed, temporary jobs distributed, trainings or trips arranged, or gifts of agricultural inputs made. The Al-Pac project, for example, has set up awards and other student programmes for aboriginal peoples in its operating area in Alberta. In Thailand, industry agents often take lucrative contracts or proposals to individuals rather than to communities as a whole, in order to fragment possible opposition.

- *Attempt to demonstrate to opponents how their concerns can be met within the industrial system other than through accepting bribes.* This strategy amounts to politely inviting opponents to accept 'translations' of their objections into the language of the industry itself, in which these objections become more tractable. Contract farming schemes, which appear to meet the common objection that pulpwood plantations are usurping individual landholdings, are one example (see chapters 12–13). Another is the promise that the suffering the expansion of plantations brings about today will be seen tomorrow as a mere 'cost' attached to the far greater 'benefits of economic development'.

- *Help see to it that resisters are crushed by force.* This option must often be chosen against local communities whose livelihoods are entirely dependent on maintaining customary powers over local commons or land, which are felt locally as moral rights. Because the logic of such livelihoods is often incompatible with that of conventional economics, such communities are often not susceptible to the last-mentioned strategy (*The Ecologist* 1993, Lohmann 1995). In particular, they are difficult to buy off because they are unwilling to accept economic 'tradeoffs' or relinquish their homes, resources or health at any price, and are often unimpressed with appeals to 'national interest'. To be a candidate for intelligent repression, however, opposition must be isolated, small-scale, poorly-coordinated, out of the public eye, or saddled with an unfavourable public image. For this option to work, too, government bureaucracies must decide it is in their own interest to foot the bill for military operations (see chapters 12–13). Thus pulp operations on Indonesia's Outer Islands are often able simply to seize thousands of hectares of the land of non-dominant ethnic groups by force, hoping that local people's isolation and fear of the government will keep them quiet (Kuroda 1995); Aracruz Celulose's initial land clearances in Brazil were also allegedly characterized by violence. In India, however, when common lands under state jurisdiction were taken over by Karnataka Pulpwood Ltd. in the 1980s, well-publicized mass marches and demonstrations during which eucalyptus saplings were uprooted and replaced with saplings of food and fodder trees resulted in an industry retreat (SPS 1989; *JV* 9–12.1990, 3.1991). A 1991–2 programme by Thailand's military government to evict millions of villagers from state forest land to make way for private sector pulpwood plantations was also stymied by high-profile, large-scale opposition.

- *Prevent certain forms of opposition from being heard by insisting on framing all discussions of plantations in the idiom of neoclassical economics and 'global demand' rather than in the languages of ordinary farmers or of everyday politics.* This strategy, an analogue of physical suppression, attempts to deny opponents the opportunity of formulating their objections in the conceptual framework of their choice. The aim is to make issues of land rights, non-economic forms of livelihood, and so on, unraisable, since industry finds it difficult to win debates framed in these terms. Thus discussion is restricted to terms on the right side of Table 6.1. The carrying out of this strategy, conveniently enough, can often be left to true-believing technocrats and other ideologues not directly tied to industry. Many academic economists and foresters, for example, even without prior consultation with industry, will shun any discussion of how paper demand is created: such discussions necessarily question their premise that human beings, as *Homo economicus*, are characterized by in principle infinitely increasing needs. Executives who know better, acknowledging the paper industry's need 'to fight for our future and create our own growth', can thus stay in the background (Clark 1994).

TABLE 6.1
Muting opposition through language

Opposition language	Industrial language
subsistence	jobs
local livelihood	national economy
power	voting, 'participation'
survival	economic growth
'death vs. life'	'compromise vs. idealism'
loss of livelihood	economic cost
land rights	economic benefits
bribes	compensation
cancelling projects	mitigating impacts

- *Acquiesce, where necessary, to certain demands made by critics.* This becomes an attractive choice where (i) opponents cannot be bought off or persuaded to modify their demands; (ii) suppression is counterproductive or impossible due to the scale, coordination, intensity, inaccessibility or public visibility of resistance; and (iii) acquiescence would not actually destroy the industry or relevant companies. Thus Western companies are slowly capitulating to strong and widespread opposition to chlorine-using industrial processes, treating it as an 'economic' reason for investing in new technology. (In doing so, in fact, they may be able to outcompete rivals with fewer resources.) Nor does the call for more recycling greatly trouble an industry long accustomed to using waste paper as raw material. Rather, it is fairly easily fed into overall supply/demand equations, and public relations officers welcome increased recycling capacity as an opportunity to claim 'green' status.

- *Move out of an area where local resistance is intractable.* This option becomes strategic in circumstances like those sketched above, but in which other regions appear more malleable to corporate intrusion. Hence it can be cost-effective for companies with sufficient global reach. Japan's paper industry, for example, has had simply to accept environmentalist resistance in western North America as an 'economic' datum beyond a political solution and accordingly shift its search for raw materials to other locations. Shell, similarly, has abandoned its pulpwood plantation plans in Thailand in order to concentrate on schemes elsewhere.

More difficult forms of resistance

Some obstacles to the expansion of the pulp and paper industry present deeper threats. No paper firm, faced with coordinated, publicly visible opposition to development of new industrial fibre sources across large areas of the globe, can buy it off everywhere it arises, smash it wholesale, or shift its search for raw materials to another planet. If, as Ronald A. Duchin of the US public relations firm Mongoven Biscoe & Duchin points out, local groups who believe that they 'should have direct power over industry' are 'difficult to deal with' (*PRW* 10–12.1993), alliances of such groups are that much more so.

Nor can any paper corporation acquiesce to calls for reduced demand for all its products. Movements which go beyond questioning one or another product (such as chlorine-bleached or non-recycled paper grades) or company toward questioning the ethos of ever-increasing per capita paper consumption itself are difficult for the industry to accommodate. The pulp and paper industry, like others, relies for its growth on the suppression of a sense of 'enough' and on the blurring of distinctions between necessity and luxury demand. The skewed distribution of consumption displayed in Tables 2.4 and 2.5 is, for paper executives, evidence not that high consumers are consuming too much but that others are consuming too little. The fact that paper demand is potentially fragile in spite of efforts to embed paper consumption ever more thoroughly into daily life — during economic recessions, for instance, advertising can fall off rapidly without consumers rising up in outrage at being deprived of magazine inserts or new four-colour packaging — makes the industry particularly defensive about critiques of infinitely-increasing demand (Price 1995). David Clark of the Confederation of European Paper Industries recently told his colleagues that the 'important question is whether paper will show the same relationship to economic development as it has in the past':

> Consumers are no longer what they used to be. . . . [they] wish to know the origins, content and method of manufacture of the paper they use . . . Their behaviour is . . . less predictable than it was. . . . Our industry stands accused, quite unjustly, of . . . creating garbage mountains. This concern . . . could . . . reduce the long-term demand for paper [as] has already happened to packaging where in most of the legislation the first priority is to use less.

. . . Many paper and board producers in the packaging sector believe that the requirement to use less and lighter packaging spells the end of growth in the sector. Similar concerns can be seen in other sectors — junk mail, unnecessary advertising, etc. Like packaging, paper itself, once synonymous with civilization and culture, could now be seen as an unnecessary and environmentally damaging material. If you think I am being unduly pessimistic and influenced by a few environmental extremists just visit any primary or secondary school to meet the next generation of consumers. Our industry can no longer afford to take long-term growth for granted. More and more we shall have to fight for our future and create our own growth. In this respect paper itself becomes increasingly a consumer product where total demand has to be stimulated. The alternative, to do nothing, could produce a static or even declining demand with serious implications for the industry, its reputation, its technology and the quality of the people it attracts. . . . Until the paper industry and its allies come together to . . . convince the public of the social and ecological value of paper, our industry will continue to be vulnerable to more and increasingly damaging environmental attacks. In this we have something to learn from other industries such as chemicals (Clark 1994).

It is a waste of time for industry to try to cultivate critics with whom it has irreconcilable conflicts concerning land or demand creation. Better by far for it to cut its losses with them and concentrate on isolating them from potential allies of different experience and social class. Unable either to crush such critics or to accommodate and conciliate them sufficiently, industry adopts the ancient strategy of divide and conquer.

That entails a dual approach. With those stubbornly resisting takeover of land or water for plantations, or the ethos of infinitely-increasing demand, industry must cut its losses and resign itself to the risks associated with showing indifference or practicing evasion, denial, repression or trickery. At the same time, it must cultivate governments and uncommitted but powerful urban and Northern middle classes, including consumers or environmentalists, who live at some distance from mills and plantations.

Thus pulp and paper interests in Indonesia have resorted to repression, abuses, and attempts to divide communities against themselves at home while hiring public relations firms to present a softer picture to customers and legislators in the West. Finland's Enso, too, while taking advantage of Indonesia's investment climate by involving itself in a joint pulpwood venture in west Kalimantan (see chapter 11), has let it be known that no less an organization than the World Wide Fund for Nature will be helping to 'assess the project's environmental impact' should any international observers have questions about the project (*PPI* 3.1995).

Jaakko Pöyry consultants adopted a similar strategy when they largely abandoned efforts to enlist local environmentalists to their side midway through the company's master planning exercise in Thailand, smearing and breaking promises to them, while at the same time helping to channel money to academics and taking advantage of the Finnish media's reluctance to publicize within Finland the nature and extent of Thai opposition to the activities of such a prominent company (see chapter 12). By the same token, when Pöyry Chief Executive Officer Henrik Ehrnrooth and top Pöyry consultant Jouko Virta were publicly criticized in Finland about Pöyry's involvement in a plantation project in the Dominican Republic, they simply denied that the firm had ever been in that country, despite being shown Dominican media clippings and photographs reporting Virta's negotiation of a Dominican plantation contract which was in fact finalized within a month (Wallgren 1994). While such economies with the truth could be expected to provoke outrage in the Dominican Republic, this does not greatly matter as long as the company's stories are believed in Finland: the tactical point is merely to prevent Dominican plantation opponents' claims from being taken seriously by the Finnish public. Similar practices are also followed in Japan. When a Kyodo News Service reporter uncovered Marubeni's illegal chipping of Bintuni Bay mangroves in Indonesia, for example, the company intervened with his boss in order to suppress publication. Japanese media coverage of the huge national outcry in Australia over wood chip exports to Japan has also been virtually nil (Kuroda 1995).

Wisely, the pulp and paper industry tends to avoid entering debates it cannot easily win, such as the debate over excessive Northern demand. Instead, it tries to prevent such debates from being heard by the wider public. One way of doing this is to attempt to marginalize opponents who try to get such debates started. Wherever 'environmentalism' has become mainstream, it has become difficult to do so by labeling them 'Communist', 'fascist', 'traitorous' or 'anti-development' as part of a strategy of suppression. Instead, industry attempts to redefine and embrace 'environmentalism' itself in a way which implicitly sets such opponents apart as radical or irresponsible (*PRW* 4.94, 1.93). Moreover, by promoting, in the North, the use of an idiom which identifies economic growth with livelihood, paper consumption with literacy, and large corporations as merely another group of 'concerned citizens', the industry strives to create an atmosphere in which grassroots resistance in the South appears cranky or, even better, inconceivable.

Arjo Wiggins Appleton executives O. Fernandez Carro and Robert A. Wilson (1992) sum up such strategies well when they urge their colleagues not to target 'apparent opposition' if that means 'forgetting the vast mass in between: the public', and not to 'respond to the mobile agenda of others' but rather to 'write the agenda and diffuse negative issues'. Politics, they go on,

> provides the packaging and the vehicle to achieve the industrial objectives. Success is measured by the freedom to plant fibre crops, recognizing the sum total of all the political forces (in the broadest sense). There are two elements to the political subsystem [of the total quality system of industrial forestry]: the message and the target. The message needs to be short, non-technical, and fundamental: for example, 'Trees are good. We need more trees not less'. Our objective should be to create and move inside an ever-increasing friendly circle of public opinion.

US public relations firm Mongoven, Biscoe & Duchin (MBD) take the counsel of Wilson and Fernandez Carro one step further by dividing environmental and other activists into four categories. 'Opportunist' activists interested in adding a string of career triumphs to their own personal curriculum vitae, MBD proposes, can be dealt with by providing them 'with at least the perception of a partial victory'. 'Idealists', who 'want a perfect world', although they may hold considerable moral authority with the public because they have no visible vested interests, can be neutralized if they can be convinced that their position is causing harm to others. So-called 'realists', meanwhile, are even easier to 'seek an arrangement' with. Because they tend to be relatively inexperienced in the workings of power outside corporate and NGO corridors, such 'realists' are easily taken in by industry's claim to be the 'only show in town'. They are also naively credulous of industry's claim that the only way of getting its attention for the purpose of effective 'damage control' is to accept its language, learn to 'live with tradeoffs', and abjure radical change. The group likely to present the most effective challenge to advancement of corporate interests, MBD concludes, are 'radicals' interested in 'social justice and political empowerment', who cannot be restricted to single technical issues.

Given this taxonomy, corporate divide-and-conquer strategy is obvious: isolate the 'radicals', cultivate the 'idealists' and educate them to be 'realists', then coopt the 'realists' into agreeing with industry. Without the support of 'idealists' and 'realists', MBD suggests, the 'radical' and 'opportunistic' positions will begin to 'look shallow and self-serving' to the public. The credibility of the 'radicals' will be lost and the 'opportunists' can then be counted on to share in the final 'policy resolution' (*PRW* 4.94).

Public relations, intelligence, and 'astroturf'

The thinking of Fernandez Carro and Wilson and MDB highlights how thoroughly industry's divide-and-conquer strategy is enmeshed in the contemporary mechanics of public relations. The future of the plantation, pulp and paper industries is likely to be highly dependent not only on advertising, the bribery of experts, and other classic forms of 'the engineering of consent', but also on intelligence-gathering, counter-intelligence, infiltration, division and funding of NGOs, cooption of environmentalists, political subversion and provocation, and the creation of fake grassroots movements. Also crucial are those related 20th-century arts of enclosing or replacing democracy known as opinion polling and cost-benefit analysis (Stauber and Rampton 1995, Carey 1995, Hitchens 1994, O'Neill 1993, Adams 1994).

Some forestry industry consulting firms carry out such activities themselves. In 1993, for instance, Jaakko Pöyry began to publish a confidential quarterly intelligence report on environmentalist thinking and activities, aimed at a clientele of wealthy companies. The editors of the report, known as *EcoDigest*, collect information partly by monitoring NGO publications, watching environmentalists' electronic mail conferences, and sending queries to environmental groups. One such query to a Canadian group was aimed at determining if 'your organization has a policy as regards forest plantations. If so, do you see them as a Good Thing (e.g., because they take pressure off primary forests) or a Bad Thing (e.g., because they reduce biodiversity)? Furthermore, what are the arguments you use?' (Ikonen 1994, Orton 1994). The Swedish Pulp and Paper Association, meanwhile, has started sending out regular 'press briefings' on the environmental virtues of tree plantations, and PaperInfo also publishes news on environmental groups in its *Environews* section.

Other important actors in pulp and paper hire outside public relations and advertising firms. Concerned about its environmental image among British paper consumers, for example, the Finnish Forest Industries Federation is paying the London firm Jackie Cooper PR to disseminate information detailing its environmental commitments; while Canada's MacMillan Bloedel has hired Pielle and Finland's United Paper Mills has retained EIBIS International to carry out similar assignments. The Finnish industry invited journalists and environmentalists on a tour of its operations in 1994 following a damaging article about Finnish forestry practices in the mass-circulation German periodical *Der Spiegel*. Indonesia's timber industry, meanwhile, has employed high-priced Western media firms, including Grey/MediaCom, to prepare advertisements claiming, falsely, that clear-cutting is not permitted in Indonesia, that areas which in fact are to be logged are being set aside as 'permanent forest', and that the country's forest management is 'sustainable'.

TABLE 6.2
Leading US PR firms working on environmental issues

Company	1992 net fees (US$million)	1993 'green' fees (US$million)
Burson-Marsteller	204	18
Shandwick	166	7
Hill & Knowlton	149	10
Omnicom PR Network	66	na
Edelman PR Worldwide	60	6
Fleishman-Hillard	59	9
Ketchum PR	46	15
Rowland Co.	44	5
Ogilvy, Adams & Rhinehart	36	3
Manning, Selvage & Lee	31	2
E. Bruce Harrison	na	7
Pacific/West Communications	na	5

Source: PRW 4.94, 1.95.

Clients of the world's largest PR company, the US firm Burson-Marsteller (B-M) — which boasts annual fees totalling over $200 million, 2,071 employees, 62 offices in 29 countries, and its own Environmental Practice Group — meanwhile include Kimberly-Clark, Asea Brown Boveri, Dow Chemical, TetraPak, Alliance for Beverage Cartons and the Environment, Shell, and the Government of Indonesia (Stauber and Rampton 1995). For more than a decade, B-M has helped forestry firms shedding jobs pioneer a novel combination of union-busting and anti-environmentalism. During a period in which it has taken B-M's advice, for example, the US firm Louisiana-Pacific has destroyed its workers' union and built pulp mills in Mexico where it has been able to pay workers less than US$2 per hour for processing US logs. At the same time, it has encouraged threatened employees to blame environmentalists for their plight, bussing them to rallies staged to counter forest protection movements and nourishing the growth of pro-logging 'grassroots' groups. After helping to sow similar anti-environmentalist sentiment across forest regions of western Canada, B-M advised its forestry industry clients to set up the British Columbia Forest Alliance, a front group 75 per cent funded by 16 forestry companies, which positioned itself as a 'moderate' organization on environmental issues (Nelson 1994). Other multinational public relations firms becoming active on the green front (see Table 6.2 for prominent US firms) can also be expected to become increasingly involved in work supporting the pulp, paper and plantation industries.

Among the services which the 'green teams' at such firms can offer the pulp and paper industry are the following:

- Lobbying domestic or foreign parliamentarians or government officials, often using parliamentary veterans, former environmentalists or former reporters, in order to secure favourable legislation or enforcement on trade, environmental, or other issues.

- Disseminating propaganda supporting industry positions, including 'scientific' evidence emanating from pundits or research organizations funded by industry, and placing advertisements and 'advertorials' in mass-circulation newspapers and other periodicals as well as commercials on television. PR disguised as news or disinterested commentary is increasingly reaching the columns of the most respectable publications and is often reproduced by mainstream or academic book publishers as well.

- Suppressing enviromentalist books before they are printed by sabotaging promotional tours, planting callers to talk shows, or convincing journalists that books are unsubstantiated (*PRW* 10.94)

- Infiltrating environmentalists' meetings in the guise of activists or housewives to gather information or 'guide' discussions. In Britain, there is considerable evidence that environmentalists' telephones are tapped both by government and by private security firms.

- Posing as journalists in order to obtain previews of research results which might be damaging to industry.

- Setting up fake 'environmentalist' NGOs with a pro-industry agenda yet no obvious financial interests (including so-called 'astroturf' grassroots groups, named after the artificial grass used in some American sports arenas) (*PRW* 10.94). Forestry corporations such as Weyerhaeuser are also creating 'community advisory panels' or conducting 'open community forums' in efforts to regulate and co-opt criticism of the industry (Weyerhaeuser 1995).

- Helping Southern elites win over Northern consumers through advertising and other means.

- Keeping clients posted on global warming, packaging, rainforests, Green political parties, seeds, pesticides, chlorine, biotechnology, eco-marketing and certification.

- Monitoring and collecting data on environmental journalists and their interests and weaknesses so that they can be better manipulated.

- Building up files on activist groups, their leaderships, methods of operation, anticipated reactions to new products, funding sources, and 'potential for industry relationship' with an eye to finding out 'what's motivating them, how serious they are, what they will consider "success"' (*PRW* 1.93, 4.94).

- Advising corporations on how to offer financial support to NGOs with heavy needs for funding and 'respectability'. Tellingly, corporate sponsors of large US-based organizations such as World Wide Fund for Nature and Environmental Defense Fund have also funded about one-quarter of the 37 groups described in the Greenpeace Guide to Anti-Environmental Organizations (*PRW* 1.95).

- Advising corporations on how to put critical individual environmentalists or former government regulators on their payrolls. One of the authors of this book, for instance, was offered a consultancy at Shell following his involvement in a campaign against Shell's involvement in a Thai plantation project.

Given the convergence of method between public relations and government intelligence or police organizations, together with the increasing interest of both in managing 'green' issues, organizations critical of the spread of pulpwood plantations can expect in the future to be the subjects of a growing variety of sophisticated political dirty tricks. Fake letters or literature will be disseminated, and incidents manufactured or provoked, which either divide environmentalists against each other or discredit them in the eyes of the public. *Agents provocateurs* placed in green movements have already been responsible for considerable violence in the US, particularly on the West Coast (Helvarg 1994), and Hill and Knowlton has distributed a phony memorandum using an Earth First letterhead calling for acts of violence 'to fuck up the mega machine' (*PRW* 1.95). Such tactics draw on the experience of the US's Federal Bureau of Investigation (FBI), which has frequently attempted to break up domestic political movements not only by murder, assault, bombings, harrassment, and subsidized book publication, but also by 'bad-jacketing' — making dissident groups suspicious that colleagues are associated with rightist organizations or corporations — and by the sowing of paranoia about state or corporate surveillance (Churchill and van der Wall 1988). In addition, the practice of cultivating public hostility toward activists by framing them for various outrages — historically used widely by Southern security apparatuses against environmentalists and others, by the FBI against US liberation and civil rights movements, and by the UK's MI5 against trade unions — is likely to be used more extensively in the future against Northern environmentalists as well.

Supporting conditions

Industry's attempts to divide plantation critics from potential allies are made easier by a number of cultural factors.

First, the idiom in which corporations prefer to have discussions framed — an idiom which tends to equate the left and right sides of Table 6.1 — is already accepted by a large part of the public to which all sides appeal. In particular, mainstream news organs such as *The Economist, The New York Times* and *Time* magazine still regularly claim that the technical 'expertise' Northern governments and corporations urge on the South is neutral, politically disinterested, and universally beneficial to livelihood. From this perspective, the anger of Southerners who see in their daily lives the damage such 'experts' can do, even on those few occasions when that anger is reported in the Northern press, often appears bewildering, irrational, even unintelligible. What seems to many Southern activists to be common sense — that 'economic development' is destroying occupations and making people hungry — can appear in the North as a contradiction in terms.

It is immensely useful to globalizing pulp and paper interests, moreover, that there exist few channels through which Finns, say, can come to understand Thai ways of life, or US citizens to understand Brazilian culture. Protests which are potent in Southern contexts thus lose much of their force when transposed to Northern ones, helping to keep Southern plantation opponents divided from potential allies in the North. Self-flattering Northern cultural lore, for example, holds there to be a predetermined sequence of developmental stages through which all societies must pass until they have replicated the Northern 'success story'. Southern plantation opponents who reject this lore are often dismissed by astounded Northern observers as being simply outside the pale. When told of Thai NGO opposition to pulp plantations in Thailand, for example, one Swedish forester, despite having no personal or institutional stake in the issue, was frankly mystified, remarking that 'as a Swede, I cannot accept that, because almost all of Sweden is covered with plantations' (Usher 1994). By disproportionately strengthening actors with a global reach, such gaps in understanding help push along the process of globalization.

Third, in many of its endeavours the plantation and pulp and paper sector enjoys the tacit support of various nominally independent experts (a few of them even associated with NGOs such as the World Resources Institute and the International Institute for Environment and Development) who are not necessarily either financially supported by industry, or particular targets of public relations firms, or otherwise directly associated with private firms. Such experts, whether foresters, biologists, or economists, tend not even to consider the possibility of making alliances with dissident villagers, even of their own nationality. One of the results, as chapter 4 has pointed out, has been the emergence of two distinct, seemingly conflicting, yet equally extensive 'libraries' detailing the effects of commercial tree plantations, one cited by proponents and one cited by critics.

Dividing experts from plantation opponents

Why do so many seemingly independent professionals often work in a way which pits them against rural dwellers adversely affected by pulpwood plantations? Part of the explanation, chapter 4 has suggested, lies with such experts' reductionist approach to knowledge. Yet to stop here is to leave many difficult questions unanswered and thus to risk political ineffectiveness. Why, for example, do many scientists, economists and NGOs who accept much of the critique of pulpwood plantations, and who are not wildly averse to putting their own fields in an interdisciplinary perspective, still act in a way which supports the abuses of the industry? What are the psychological, cultural, and political roots of their stance? If 'reductionism' is a problem, can plantation critics do anything to encourage such intellectuals to abandon it? The loyalty which many experts hold for the idea that it is mainly industry and capitalism which hold the answers to the problems they create, in short, is of a type and degree which requires an explanation which has not yet been given.

Many experts, of course, frustrated by having to work for organizations which they know cannot act on their most important recommendations, have learned, as a matter of psychological self-defence, to moderate those recommendations while averting their gaze from the political meaning of such actions. Some are even driven into denying that their actions *have* a political meaning. As one eminent Yale University forester has remarked, 'There are so many foresters accustomed to taking orders from on high that they don't question. They just go on drawing their pay and saying it's not their fault' (Vail 1993). Another part of the explanation of the limitations and lacunae in the thinking of many scientists, economists, and NGO staff lies in the class, social, or educational background which they tend to share with each other and with industry figures (Dove 1992, Frossard 1995). This tempts them to seek out, and listen to, each other and to industry, rather than delve too deeply into unfamiliar facts or social milieux. The simple lack of awareness on the part of many experts of the many existing social alternatives to 'working with industry' should never be underestimated. The familiar notion that certain scientific 'paradigms' exercise a hegemony over experts' imaginations also helps explain the pervasiveness of pro-industry attitudes among intellectuals who seemingly have no vested interests — although, unless this explanation is combined with a thoroughgoing sociology and history of scientific practices, it leaves mysterious where these 'paradigms' come from and how they might be challenged.

A more fundamental part of the explanation perhaps has to do with many experts' innocent beliefs about power. In the academic, planning, and development circles in which many experts and large, well-financed NGOs move — circles which have flourished during the process of globalization — power is frequently represented as something which industry and its governmental and intergovernmental allies 'possess' and which others (who are characterized as their beneficiaries or victims) 'lack'. It is not difficult to see what nourishes this conception. Experts are accustomed to the interplay of power in the restricted arena of their offices, laboratories, and social milieux, where experience teaches them that industry, the state and international organizations 'get things done'. The easiest way such experts have of conceptualizing other actors with whom they are less familiar is to assume that they are what these bodies are not: that is, powerless.

Professionals have a vested interest in viewing themselves, moreover, as repositories of knowledge which others lack. Bringing about the changes they desire, they often assume, must then be a matter of ferrying that knowledge to industry, the state and their allies, who, as mentioned above, are in turn seen as having the power that others lack. This does not mean that experts always agree with industry and the state, or never take the side of their victims. It does mean, however, that if experts do fight for the downtrodden, they often are willing to do so only if they are allowed to present their case in lawyerlike fashion, and transformed into proper 'knowledge', to the people they imagine to be 'the powerful'. The most important audience for the truth, the assumption goes, is industry, the state, and their allies.

Of course, as experts are the first to admit, 'speaking truth to power' in this way may not always get results, despite the flattering noises industry makes to court expert endorsement. To have the chance to speak truth to power even once in a while, moreover, requires staying on power's good side and thus acceding to frequent and frustrating compromises. But making compromises with industry and the state, the reasoning goes, is at least more likely to result in effective 'damage control' than making compromises with impotent villagers. Surely, hopeful professionals reason, governments, corporations and international agencies are likely to let us be their 'brains' at least some of the time!

It is only on this naïve, dichotomous have/lack view of power and knowledge that some of the preposterous utterances of experts who have formed *de facto* alliances with industry begin to make sense. The claim of one IIED consultant that to reject the framework of the industry-oriented Tropical Forestry Action Programme would be to 'start from square one', for example, is intelligible only on the assumption that opponents of the plan are so powerless that supporting their own environmental movements would be a dead end (Sargent 1990b). An Audubon Society official's insistence that 'conservationists have just got to learn to work with industry', similarly, is plausible only if one excludes from the outset the possibility that one can influence the establishment game by not playing it as well as by playing it (*PRW* 1.95, Ferguson 1994, Scott 1990). The notion that good environmental practice or local communities' rights can be 'secured' only by appealing to governments or to the United Nations is based on the same illusory dichotomy between official power and vernacular impotence, as is the widespread notion that the only way to engage in effective environmental 'damage control' is to participate in activities dominated by states, international organizations, and business.

Indeed, it is probably not too much to say that the have/lack view of power and knowledge — and the Orientalism and lack of intercultural engagement out of which this view grows (Said 1978, Lohmann 1993) — lie at the root of a whole cluster of false clichés about social change.

These include the cliché that activists must choose between, on the one hand, acting on principle and being impotent and, on the other, compromising and being effective; between having 'clean hands and living in an ivory tower' and having 'dirty hands and living in the real world'; between being uncompromising but powerless 'idealists' and compromising but influential 'realists'. Only on the odd assumption that ordinary people are not engaged in power politics, and do not dirty their hands with compromises, could such ideas get off the ground. Only on the assumption that effective grassroots movements do not exist could the alternative to 'working with industry' be seen to be 'living in an ivory tower with only one's principles for company'. And only on the assumption that governments or international organizations are powerful but neutral machines awaiting programming by experts could those experts allow themselves to be divided so easily from potential allies at the grassroots.

So seductive is the image of power as something which the state, international organizations, and industry 'have' and others 'lack' that it often prevents evidence of grassroots power from even being seen. Has a plantation scheme been cancelled in a Southern country? Has a logging ban been instituted? Has a pulp mill improved its effluent treatment? The natural assumption of many experts is that such things can't have happened because of popular resistance. There must be some other, invisible explanation. Some Western environmentalist must have met quietly with officials to point out the error of their ways. Or perhaps a secret deal was struck at high governmental levels to get logs from some other country. Or perhaps a think tank released a new cost-benefit analysis or proposed a new natural resource accounting scheme. It thus becomes difficult even in principle for many experts to recognize counterexamples to their assumption that there is only one true language of power, that spoken by themselves, business and officialdom. While this assumption flatters professionals' feelings of self-worth and helps expert institutions secure funding, it only further entrenches contempt for the public.

Ironically, the have/lack picture of power, regarded as so 'realistic' by its acolytes, is one to which corporations and governments themselves — at least those which have lasted — have never subscribed, although it is in their interest for them often to advertise themselves as doing so. As this book makes clear in nearly every chapter, when the plantation, pulp and paper industry speaks to itself, or maneuvers to gain 'freedom to plant', or maps strategy with PR specialists, it reveals a persistent and pragmatic preoccupation with grassroots resistance and the opinions of ordinary people. While the industry is also concerned to win over the newspaper, the government official, the college professor and the professionalized NGO, fear of the irate crowd who may be so stupid or disobedient as not to listen to these luminaries is never far from its collective mind. As always, the self-styled 'realists' eager to 'speak truth to power' turn out to be largely the victims of their own naïve and self-aggrandizing fantasies, encouraged by the genuine realists of the corporate and official worlds who, by contrast, are always conscious of having to act against a constant background of opposition. The more that potential professional critics of plantations are neutralized in this way, the more easily effective expert–grassroots political alliances can be prevented.

Stories for the uninformed public

Most plantation industry propaganda is disseminated as part of the general divide-and-conquer strategy for managing difficult opposition. This propaganda seldom seeks to convince critics who already have a good knowledge of local plantation conditions or of industry maneuverings; as Fernandez Carro and Wilson intimate, these critics are not likely to be susceptible to it. Rather, industry propaganda tends to be aimed at consumers, officials, and environmentalists whose day-to-day lives are not touched by production operations, and who have little time to acquaint themselves with the technical issues. Correspondingly, it is usually deployed through the channels which will reach these groups best: public debates, newspaper columns, and governmental negotiations. The purpose throughout is to prevent critics with some knowledge of the industry's actions from finding support from potential allies in other circles.

Among the bite-sized 'messages' which the plantation and pulp and paper industry have disseminated to this end are the following:

• *Pulpwood plantations can 'relieve some of the pressure' on the world's natural forests* (Bazett 1993: 100).

Without extensive plantations being established in the South, it is said, and with continuing rising demand, both local old-growth forests and distant ones such as those in Siberia will become more economically attractive to the pulp industry; yet with plantations they can be saved. This 'message' has a theoretical attractiveness which can appeal to Northern groups not directly acquainted with the facts on the ground, but must be used with care elsewhere. For one thing, as the case studies of Part Two will show in detail, for historical, technical, economic and political reasons alike, plantations and increased logging of natural forests often go hand in hand. The logging of natural forest often provides the necessary funding for, or is justified by, the establishment of industrial plantations. In Chile, for example, many plantations are being established by logging native hardwoods which might otherwise have survived; as one industry source admits, the sudden expansion of wood chip exports has 'exerted pressure on native forests, even though the pulp and paper industry uses only plantation wood' [sic]. In Argentina, a British-based company planned to embark on an integrated scheme to log in a 50,000-hectare area containing natural quebracho (*Schinopis lorentzii*) and algarrobo (*Prosopis* spp.) forest in Santiago del Estero and then replace it with eucalyptus, until environmentalists got wind of the scheme. In Brazil, Veracruz Florestal has destroyed native forests in Mata Atlantica in Bahia to plant eucalyptus, with advice from Jaakko Pöyry. In Thailand's Nakorn Ratchasima and Chachoengsao provinces, too, as well as in Malaysia, Kalimantan and Sumatra, logging and plantations have gone hand in hand. Many plantations in Indonesia are subsidized by logging revenues recycled through the state. Second, the establishment of plantations on already deforested land tends to induce the inhabitants to settle in forests elsewhere (as has happened in Thailand), resulting in deforestation at a distance, and sometimes even setting off a chain reaction of further impoverishment and forest destruction. (See Part Two.) Finally, pulpwood plantations are not designed to supply industrial sawlogs, nor logs for plywood, and so can hardly remove what are probably the dominant motivations for logging tropical forests, particularly in Southeast Asia and the South Pacific.

- *Pulpwood plantations can increase aggregate tree cover.*

'Trees are good,' the industry rationale goes. 'We need more trees, not less.' Again, this disingenuous 'message' is plausible but must not be used with audiences aware that large monoculture pulpwood plantations are, as suggested above, merely a continuation of deforestation by other means. It must also not be used with audiences sensitive to the distinction between mere tree cover and forest. The exotic, homogeneous trees of an industrial plantation cannot fulfil the environmental functions of the trees of a forest, which interact with soils, the water cycle, animals, other plants, and surrounding ecosystems in a way that assures the regeneration potential of each.

Tellingly, the confusion between forests and plantations which the industry is eager to promote with uninformed audiences is one which it is equally eager to disavow with more sophisticated audiences. With the latter, the industry insists that pulpwood plantations be judged not as forests but as agricultural crops (Aracruz Celulose 1996), asking its critics why they do not condemn fields of maize with equal fervour. Needing to appeal to audiences with different levels of understanding, the industry must try to have it both ways.

- *Indefinitely rising paper demand is either inevitable or desirable or both.*

This assumption is so fundamental a part of the contemporary ideology of economic growth that it is seldom stated explicitly and even less often discussed. To call it into question among a business audience tends to elicit either silence or the rebuttal 'Do you want to go back to the Stone Age?' It is reinforced in other arenas by uncritical reference, to, for example, FAO projections of demand growth. Combined with the idea that further exploitation of old growth is undesirable, the assumption leads naturally to the conclusion that plantations must be expanded and that side effects, though regrettable, are unavoidable. The search for 'alternatives' to the current paper and pulp production model, instead of being concerned with such social matters as demand creation or the desirability of junk mail or mini-packaging, thus becomes narrowed to the search for 'alternative technical means of increasing production'. For an appropriate response to this ideology, it is useful to turn once again to the late ex-FAO forester Jack Westoby:

> Can we be sure that it is in the public interest that pulp and paper should be a growth industry? . . . [I]t irks me to have to carry about a kilogramme of newsprint when all I want is a gramme of news. I also begrudge the time it takes me to transfer 50 or 60 unsolicited circulars from my letter-box to my waste bin every week . . . My blood pressure rises each time I pick up a package at the supermarket and discover when I get home that it takes at least ten minutes and a good deal of ingenuity to penetrate it and reach the contents. The fact is that the pulp and paper industry produces a lot of things I don't want, that I never asked for, but that I am compelled to have and also compelled to pay for, directly or indirectly. As an individual, I resent all these things. As a member of the community, it strikes me as irrational.

Westoby adds that if a country 'sets as its goal a specific future level of GNP per head, and assumes that this carries with it a connotation of so many kilos per head of paper and board, . . . it is setting out on the wrong foot'. Treating such numbers as divinities is like projecting recent US heroin-consumption trends in linear fashion into the future, which rapidly leads one to the conclusion that 'every man, woman and child in the US will be a junkie by 2020' (Westoby 1987).

- *Demand for paper comes not from particular groups, classes, or societies, but rather from 'the globe' or 'the nation' as a whole.*

This notion, too, is seldom made explicit, but is a natural one for many participants in the plantations debate. For forestry businesses intent on expanding in the South, it is useful because it helps obscure exactly who paper consumers are and thus helps erode the common sense observation that a huge demand in, say, Japan or the US does not necessarily justify plantation expansion in Southeast Asia. Thus a Jaakko Pöyry executive arguing for a plan promoting plantations in Thailand may repeatedly refer to the necessity of 'meeting demand' while carefully leaving unspecified whose demand is being referred to (Kärkkäinen 1994). For academic economists, UN statisticians and researchers at organizations such as World Resources Institute or WorldWatch, on the other hand, treating 'demand' as global is sometimes perhaps merely an analytical convenience in an age in which the paper market has spread to all parts of the planet. Whatever the intent with which it is used, however, the construction of the category of 'global demand' has a number of deep practical effects. By suggesting there is a type of need which, being 'global', overrides merely 'local' claims to land and water, the idea helps justify siting plantations at any location on the planet that industry deems appropriate. Similarly, it elevates officials and businesspeople who advertise themselves as addressing 'global' concerns into a superior moral position. It becomes permissible to assign local rights and social structures to the category of 'eggs' which may be 'broken' in order to make the 'omelet' which satisfies an exalted 'global' demand, around which national and local forest management, forest industries and state institutions are to be adjusted. The notion of 'global demand' also licenses cross-regional and cross-class subsidies from the poor for commercial and export initiatives, as well as large forced evictions; the disdain for local practices, local knowledge and local conservation which results from centralization; and the setting of numerical global goals for plantation area. Looking at variations in demand both among countries (see Table 2.4) and within countries helps to call into question such practices.

- *Pulpwood plantations are an economically productive use of unoccupied, degraded land which would otherwise rest unused.*

This 'message' is used across the entire South both by industry and by government. It can be effective with environmentalists unaware of industry thinking and practice at the grassroots, but tends to fall flat with those acquainted with the facts on the ground. As the Asian Development Bank, a Shell International researcher, and Shell (Thailand) have all pointed out, industry is not particularly interested in degraded land. What is required for competitiveness, rather, is contiguous chunks of 'land suitable for superior biological growth rates for those species the market wants' as well as 'year-round water' and easy access to nearby processors or ports — one example being Shell's own ill-fated project area on the well-watered (with 1800–2500 millimetres of rainfall per year), relatively fertile fruit-growing soils of Chanthaburi, eastern Thailand (Bazett 1993, Apichai 1992). In India, industry has consistently refused to take degraded land offered for its use, preferring good-quality lands on which the poor are dependent for their livelihoods (*DTE* [Delhi] 31.8.1995, Guha 1988). Also attractive, in many countries, is densely-forested land (see chapters 7, 8, and 11). As an Indah Kiat executive has stated forthrightly, 'Basically we are looking for forest which can be clear cut and replaced with eucalyptus and acacia' (cited in Marchak 1992). To quote the Asian Development Bank on Indonesia,

> . . . while there is no doubt that there is sufficient degraded and unproductive forest land to provide the postulated areas, there is little evidence that it is in appropriate locations and there is every possibility that pressure may be brought to bear to release productive forest land for clear-cutting and replanting with fast-growing exotic species to satisfy the perceived needs of the pulpmill proposals. Even where there is sufficient degraded land, the use of high quality production forest is likely to show more attractive rates of return and more certain financial viability than successive rotations of fast-growing species on difficult and impoverished sites (ADB 1993).

This is confirmed by Philip M. Fearnside (1993), a leading expert on the forests of Brazil, who notes that

> plantation projects on degraded lands are inherently less attractive than cutting down forest, because there is no free wood from the native forest and the degraded soil leads to slower growth of the planted trees. This is very important to firms that might want to plant trees, as the financial returns from plantations are highly sensitive to the growth rate of the trees. Because it costs approximately the same to plant trees on a site with poor soil as on a site with good soil, and most of the costs are in . . . planting, harvesting, and maintaining the trees for the first portion of the growth cycle, the profit margin changes much more than the growth rate changes. If the trees grow ten per cent faster, the profits may be, say, 50 per cent higher.

Individual farmers, too, often wind up planting eucalyptus on highly-productive land when the market for other cash crops is faltering and labour is short, as has happened in India in western Uttar Pradesh, Haryana, Punjab and south Gujarat (Saxena 1992a). Finally, it must be remembered that what counts as 'degraded' or 'unused' land depends entirely on who is talking. Land identified in official

Geographic Information Systems as 'marginal' or 'unoccupied', for example, may in fact be used by local people as woodland commons, pasture, or fallow for rotational agriculture.

- *Plantation expansion helps make underdeveloped countries 'self-sufficient' in paper.*

This 'message' has been occasionally used by, for example, industry executives attempting to justify plantation development to socially-concerned audiences (Kärkkäinen 1994, RFD 1993). It is unlikely to be successful, however, with audiences aware of the actual economic and political imperatives driving the growth of plantations in the South. Such audiences will be aware, for example, that Brazil's, Chile's, Congo's, Swaziland's and Indonesia's plantations or pulp capacity are aimed largely at export; that Thailand is already self-sufficient in paper in aggregate terms and, in those sectors in which it is not, new Thai plantations will not help make it so; and that nominal self-sufficiency counts for little in the face of the liberal trade policies advocated by the industry itself, which will push pulp and paper imports into any country not producing them more cheaply.

- *To argue against the large-scale spread of pulpwood plantations is to deny Third World schoolchildren the books they need, or to claim that only Northerners are entitled to milk cartons.*

This sentimental message, when floated by industry executives in public forums, cynically hopes to trade on its audience's ignorance. As the industry knows perhaps better than anyone else, new plantations are established principally to feed surges in consumption in heavily-consuming industrial and elite sectors — particularly for items such as advertising and export and supermarket packaging — and do not respond effectively to the cry of the downtrodden for education, food, and basic necessities. (See chapter 3.) The message also quietly begs the question of what the outcome would be of potential negotiations between the downtrodden resisting the establishment of plantations around their homes and others, downtrodden or not, who want to use paper derived from those plantations. Why, after all, should it be treated as a foregone conclusion that paper consumers would not change their expectations if they had an opportunity of entering into a dialogue with those affected by the plantations which feed them? Nevertheless, the message often powerfully sways socially-concerned outsiders who are unversed in the structure of paper demand or the undemocratic way in which it is created. For example, the Executive Director of one prominent Northern NGO recently remarked dismissively, echoing mainstream economic ideology, that to investigate paper demand critically would involve dictating to ordinary people what their wants and preferences should be (Sandbrook 1995).

- *Fast-growing monoculture plantations are up to ten times more productive than natural forests.*

In one sense of 'productive' — 'productive of trees with market value as pulpwood over at least two or three growing cycles' — this 'message' is true. In other senses of 'productive', however — for example, 'productive of other trees, of animals, vegetables, fruit, mushrooms, fodder, fertilizers, water retention capacity and medicines, and of improved agriculture on neighbouring fields' — it obviously is not. Hence this 'message' cannot count in favour of plantations unless it is accompanied by an explanation of why the first sense of 'productive' should be privileged over others. To be democratic, this discussion must include affected people speaking in their own idioms. Yet when used by industrialists, state foresters, and some economists, the claim that plantations are 'productive' is often used in a context in which other meanings of 'productive' are ignored or suppressed and those who might articulate them excluded from the discussion. In such contexts, the claim is misleading and the discussion biased.

- *Fast-growing tree plantations can help curb global warming.*

This notion has proved extremely attractive to Northern elites who are reluctant to cut their fossil-fuel emissions, disinclined to master the intricacies of improving energy-use efficiency, and naïvely confident of their corporations' ability to implant large-scale plantations in any political context in the world. It has already legitimized environmentally-damaging plantations in Guatemala, Ecuador, Bolivia, Peru, Costa Rica, Brazil and Malaysia — most of which are intended to 'offset' carbon dioxide emissions from Northern industries. Various proposals have also been presented to the Global Environment Facility of the World Bank (GEF) to establish carbon plantations in Africa with co-financing from electric power firms in the North (Fearnside 1993).

Some industrialized-country environmentalists such as Norman Myers have even gone on record with the preposterous claim that grand-scale plantations are *the* leading way of buying time in which to cut CO_2 emissions (Myers 1991). The Noordwijk Declaration, signed by 63 countries in 1989, reinforced this wild notion, calling for an increase in the rate of tree plantation establishment to 12 million hectares annually by the year 2000 as a response to global warming. Corporations and multilateral and bilateral agencies, too, have learned to wave their hands about 'global warming benefits' when attempting to justify plantation projects they support for other reasons. For example, a recent report sponsored by the Swedish Forest Industry Water and Air Pollution Research Foundation recently found that 'Swedish forests bind greenhouse gases' even when the effects of pulp, paper and timber production using those 'forests' as raw material were taken into account (*Skogsindustrierna* press release 29.4.1994).

However, such arguments are not very effective as justifications for industrial plantations with audiences aware that in any economic and political regime in which large fast-growing plantations are likely to be established, it is far more probable that they will be a *cause of*, than that they will be a *solution to*, global warming. These audiences will be conscious of plantations' powerful and difficult-to-calculate role in accelerating deforestation, whether through direct encroachment on forests or through the displacement and deskilling of local people (see chapters 7, 8, 11 and 12). They will also point out that the peak carbon mass of a plantation is far less than that of a natural tropical forest. For instance, as chapter 7 notes, the above-ground biomass of *Gmelina* plantations ready for harvest at Jari in Brazil was only a quarter that of the native forest they had replaced (Russell 1983), making nonsense of official claims that the plantations helped alleviate global warming.

A 'message' about the supposed global-warming benefits of plantations is also unlikely to impress audiences who know that plantations are capable of sequestering carbon only temporarily. Even the longest-lasting wood products are unlikely to be able to keep their carbon out of the atmosphere for more than a few decades, while paper made from plantations may well be oxidized within months of harvest.

Yet even if plantations could be fashioned that resulted in net sequestration of carbon, and even if they were capable of sequestering it for a meaningful length of time, the sheer scale of the plantations that would be required makes the idea politically and technically impracticable. According to one estimate, reducing atmospheric CO_2 to preindustrial levels would require a plantation capable of producing annually many times the expected global demand for wood (Grainger 1990), or a tree farm 15–20 times the size of Britain. The fact that growth rates in large-scale plantations are always much lower than in test plots suggests that even this may be an underestimate.

Proposing industrial plantations as a meaningful response to global warming, in short, demonstrates a lack of perspective — particularly given the multitude of more feasible approaches to the problem which are available, including reductions in fossil fuel use and no-cost energy efficiency measures (Rocky Mountain Institute 1990). As even Philip Fearnside (1993), who has some sympathy with the idea of plantations as a part of a solution to global warming, notes, 'plantations are a much less cost-effective means of avoiding net emissions of greenhouse gases than is reduction of tropical deforestation, as well as having many fewer benefits in realms other than global warming abatement'.

The attempt to justify large plantations in the South on the grounds that they will alleviate global warming, finally, will be especially ineffective with audiences who question why the South (or, for that matter, poor rural regions of the North) should provide ever-expanding sinks for infinitely-growing industrial carbon dioxide sources in industrialized regions (Barnett 1992, Sargent and Bass 1992, Shell/ WWF 1993). Any historically- and scientifically-informed approach to the issue will have to acknowledge that global warming is primarily due to the industrial activities of the North, and that that is where the burden of the main changes necessary for its alleviation must lie.

It is perhaps worth adding that while plantations whose harvest was prohibited might be temporarily useful in mitigating global warming, these are not the sorts of plantations advocated by the pulp and paper industry. In any case, such plantations would eventually cease to be carbon sinks, achieving a rough carbon equilibrium with the atmosphere. Faced with such considerations, some scientists have proposed burying huge masses of logs under ocean sediments after harvesting them, sequestering their carbon permanently, and then replanting the ground where they had grown, establishing a perpetual carbon sink. Although the elaboration of such techno-fantasies undoubtedly promises years of amusement for boffins, it is unlikely to appeal to pulp and paper industrialists, who would like to use scarce plantation land for quite other purposes.

- *Plantations are profitable to local and national economies.*

This message, which is frequently floated by governments eager to attract plantation investment, is most effectively used with audiences unaware of the massive subsidies channeled to the industry mentioned in chapter 5. It would be of little use with a Uruguayan audience (say) which was aware that in that country, government subsidies underwrite roughly half of the expense of establishing pulp plantations, which would be commercially unviable without them.

- *Plantations create employment, either directly or indirectly (through transport, pulp and paper industries, and sales).*

This argument is also standard among governments promoting plantations, although industry and its consultants are less likely to resort to it now than formerly (Shell/WWF 1993). As chapter 4 shows, the claim is false. Large plantations generate direct employment mostly at planting and harvest times. In Asia especially, they tend simultaneously to deprive previous occupants of the land of their former occupations, so net employment trends may well be negative even at this stage. After planting, employment drops off substantially. Growing mechanization is reducing harvest-time jobs even further. In general terms, plantations create much less employment than agriculture, the balance being positive only in extensive, sparsely-populated livestock-raising areas. As for industrial employment, plantations do not always result in the creation of local industries, as in many cases production is aimed at the direct exporting of unprocessed logs. Even where pulp and paper industries are set up, the high degree of mechanization implies few employment opportunities, as chapter 2 reveals.

• *State-of-the-art pulpwood plantations exemplify forestry's Green Revolution in harnessing science and technology to the goal of bringing more people a better life.*

This argument, which has been put forward by some corporate technicians, is likely to attract, say, *Economist* leader-writers rather more than many of those who were at the receiving end of the original agricultural Green Revolution. The Green Revolution succeeded above all in making profits for big business, especially Northern business, who benefited from sales of seeds, fertilizers, pesticides, tractors, and so forth, but devastated many existing ecosystems and societies (George 1988, Shiva 1991a). In some ways, the effects of the plantation boom promise to be even more biased than those of the Green Revolution, in that it is focused even more tightly on promoting exports of a cheap raw material — in this case wood — at the expense of local peoples and their land. The 'scientific management' which is advertised as a benefit of plantations in reality consists of the ability to produce fibre in the shortest space of time and as cheaply as possible, while social and environmental impacts tend to be externalized.

• *Promulgating plantation guidelines is the crucial step required to make plantations 'sustainable'; if these principles can be agreed upon by all interested parties, then abuses should be avoidable through a few adjustments to production systems.*

This extremely subtle assumption is implicit in, for example, the guidelines formulated by Shell/WWF, the International Timber Trade Organization, the Canadian Pulp and Paper Round Table, the New Zealand Development Assistance Division and the Forestry Stewardship Council, as well as in most eco-labeling schemes (Shell/WWF 1993, Clark 1994, IIED 1995). It is also easily embraced by a wide variety of Northern academics, scientists, technocrats, public relations executives and environmentalists whose history, institutions, and jobs give them incentives for believing that if theory, objectives, or sound legal measures can be formulated 'correctly', even by those who live far from plantation areas, then good practice, implementation or enforcement will follow fairly straightforwardly through the efforts of existing effective, disinterested, benevolent institutions.

Such figures, when interrogated by critics with experience of some of the disastrous effects plantations can bring about in practice, often consider it sufficient simply to disavow responsibility, protesting (for example), 'But that's not the way it's supposed to work in theory', or 'The implementing agencies didn't follow my instructions'. Whether the agencies or social structures within which such ideologists and scientists work have any interest in or capability of acting on those instructions is frequently considered to be 'not my department'.

Thus one British advertising executive recently accused of making misleading claims for Indonesian forestry in a television advertisement retorted that what actually happens in practice in Indonesia is irrelevant. Evidence of rampant industry deforestation in Indonesia should not disqualify advertisers from saying that forest use there is sustainable, he maintained, any more than the evidence for the existence of drunk drivers in Britain should prevent anyone from saying that drink driving is not allowed there (Brooks 1994). In the same vein, Philip G. Adlard of the Oxford Forestry Institute has responded defensively to documentation of cases in which eucalyptus planting has reduced food production by invoking an irrelevant Cloudcuckoo-land of contextless, want-satisfying markets, uncoerced economic actors, and benevolent state intervention:

> If market forces are allowed free play this would result in a rise in price of the local staple foods and a reversal of the trend to grow tree crops for cash benefits in place of food crops needed by the local community. The market price for forest produce (poles and pulpwood) would also lead to checks on the areas planted ... The farmer has the right to choose whether he grows a cash crop and buys his staple food or grows some or all of his food crop and foregoes possible cash benefits. If market forces do not lead to a balance, or if there are proved ecological reasons for not growing a particular crop, then some form of subsidy may be the best incentive to good land use (Adlard 1993).

Part of what makes such responses so inadequate is the fact that, even in the unlikely case in which principles of plantation management could be formulated which, on paper, respect the claims and desires of people in plantation areas, such principles by themselves turn few cogs in the economic, political and social mechanisms of contemporary industrial capitalism. A number of immense practical difficulties immediately intervene — all of which, however, tend to be more familiar to activists on the ground than to office- or trial-site-bound environmentalists or technocrats.

One problem is that implementing, enforcing and financing institutions are never 'neutral' and tend either to ignore or to reinterpret management principles as they see fit. Nor do idealized egalitarian social structures of the type Adlard invokes actually exist in the real world. In 1989–90, for example, Shell hired the International Institute for Environment and Development to formulate management recommendations for a proposed plantation operation in Thailand to ensure the project was carried out in a socially and environmentally responsible fashion. Yet, as was obvious to Thai observers from the beginning, there was never any chance that IIED's plan, even if formulated with the best will in the world by the foreign experts responsible for it, could engage with Thai political and social realities in the way advertised, or indeed do more than shore up repressive forces while providing a cover for business as usual. A Shell (Thailand) executive charged with implementing the project inadvertently summed up the difficulty when he stated forthrightly that the company would follow those recommendations which were 'consistent with profits'.

Similarly, none of the environmental controls recommended by Papua New Guinea government consultants for Japanese corporate pulpwood clearfellings in the Gogol Valley — which included restrictions on logging coupe size and measures to alternate clearcuts with lightly-logged forest patches — were ever implemented (Lamb 1992). In a parallel case, Jaakko Pöyry's *Forest Policy*, although explicitly committing the company to maintaining species biodiversity and to advocating that 'any natural forest area which demonstrates untouched unique ecosystems be set aside for conservation even it it has been assigned for industrial forestry', has not prevented the firm from involving itself in (for instance) several enormous projects in Indonesia which are explicitly designed to start up by feeding off mixed tropical hardwoods from native forests (Pöyry n.d. e). Multilateral development banks' structural inability to abide by their own guidelines has also been exhaustively documented (Rich 1994, Ferguson 1994).

A second, equally deep problem with promoting plantation guidelines as a solution to plantation problems without investigating their political context concerns information and monitoring. To take one indicative example, B&Q, a large British do-it-yourself outlet which has come under heavy environmentalist pressure due to its sales of tropical timber, has.hired staff to audit its wood supplies to ensure they meet rigourous standards of 'sustainability'. Yet in a highly-extended global commercial system, with its myriad, convoluted and difficult-to-trace long-distance transactions among agents many of whom have some vested interest in concealment or prevarication, it has proved hugely difficult in practice to guarantee a 'sustainable' pedigree for many wood products, no matter how well-formulated the standards are (Cox 1993). Even more fundamentally, any team of experts sent out to certify that plantation operations in the South are abiding by management principles, because it is likely to consist of professionals of a certain class and background, will communicate far more easily and comfortably with corporate and state personnel than with affected rural residents, to understand whose observations might indeed require months or years of cultural immersion. Such experts may not find time to meet even briefly with local people requesting opportunities to pass on critical information, as happened in 1989 in Sarawak when the noted ecologist and conservationist Lord Cranbrook was enlisted by the International Timber Trade Organization to review Malaysian logging practices. The report any such 'certifiers' file after a brief visit will thus inevitably be biased and riddled with gaps.

Participating uncritically, with industry or with others, in efforts to formulate principles or guidelines for plantation management, therefore, runs the risk of merely helping corporate efforts to delay structural change or to provide ideological cover for continued depredation. This is not to deny that, on some occasions, formulating such principles can be one step in a series of actions and arguments which effectively challenge industry practice. Framing principles prohibiting wood from any extensive industrial exotic monoculture from being certified as 'sustainably produced', for example, could be part of a realistic programme to reduce the damage done by such plantations. Nor is there any point in denying that the exercise of formulating such principles can help environmental groups clarify their own viewpoints to themselves. The question of whether negotiating over the content of such principles is a waste of time or not, however, is an empirical one, to be settled by a close and, if necessary, anthropological examination of the social and political context — including the context of industrial structure and economic centralization — in which the principles are expected to operate. Such an examination, to be meaningful, must be conducted *before* negotiations about the principles' content begin, rather than after they are completed.

Conclusion

The 'messages' and assumptions described in this section, deployed selectively and shrewdly, foster the globalization of the pulp and paper industry by helping block alliances between grassroots groups fighting monoculture pulpwood plantations and environmental groups elsewhere. Yet the converse is also true. It is only the global reach of the contemporary pulp and paper industry — its ability to exploit the spatial and cultural distance between grassroots movements and intelligentsias elsewhere — that allow it to deploy its mystifications — 'Trees are good. We need more trees not less' — to drum up support for industrial tree plantations among largely urban and Northern power bases. This support is crucial, since a ballooning 'free market' in wood fibre, pulp and paper can be constructed and coordinated only if the subsidies given to consultants, foresters, aid agencies, and non-governmental organizations to promote plantations can be justified before a large and diffuse public.

To use such mystifications, however, is always to gamble that they will not be exposed. Whether that gamble pays off — whether the pulp and paper industry gains its 'freedom to plant' at the expense of the people directly affected — depends largely on the skill in intercultural conversation of plantation opponents. The prospects are far from hopeless. In a world thronged with naked emperors, paper industry figures claiming sustainable environmental benefits for large-scale monoculture pulpwood plantations are more notable than most for their sartorial minimalism.

Part Two
Tree Plantations in the South

Chapter 7

Brazil: The Eucalyptus Pulp Giant

Many people think of Brazil as an enormous, half-empty country, where the impact of a few hundred thousand hectares of eucalyptus must be minimal. This impression is mistaken. As in other countries, eucalyptus plantations in Brazil are not located on land no one wants, but rather are concentrated in the most populated coastal zones, close to the ports from which pulp is exported abroad. Plantations and pulp mills can be found from the extreme southern province of Rio Grande do Sul to Pará and Amapá in the extreme north. They have a great impact not only on land, vegetation and water, but also — more importantly for us — on people and their livelihoods.

Brazil has had long experience in planting eucalyptus for charcoal production and conifers for pulp. Indeed, the country's largest plantation area is still occupied by eucalyptus charcoal plantations. As late as the mid-1960s, however, the area planted with exotic species was relatively small and concentrated mainly in the state of São Paulo, where wood consumption was highest and where deforestation had necessitated costly imports of raw materials from elsewhere in Brazil.

The situation changed radically after a military dictatorship came to power in 1964. New policies favouring large-scale capital were implemented, forming part of a modernization strategy seeking to gain the Brazilian economy access to the international scene (Gonçalves 1992). A new forestry code was also developed, and fiscal incentives were given to promote tree plantations. Large plantations quickly sprang up in many regions of the country, in many cases without the developers having much idea of what the wood planted would finally be used for (Knight 1991b). Poor location and poor maintenance led to the loss of many plantations. Of the six million hectares planted under the government incentive system, only 2.3 million hectares of eucalyptus and 1.6 million hectares of pines survived (Bazett 1993). Years later, however, the pulp industry was able to take advantage of these extensive sources of raw materials — although to what extent is unclear, since in 1994 the total pulp plantation area was estimated by industry sources at only between 1.2 and 1.5 million hectares (three-fifths of which was eucalyptus) (Associação Nacional dos Fabricantes de Papel e Celulose n.d., Higgs 1994).

Despite the apparent anarchy in the plantation sector, the military government clearly aimed to establish a large export pulp industry. A late-1960s package of tax breaks promoting pulping machinery imports was followed by National Paper and Pulp Plans I (1974) and II (1979/80). The industry became one of the main beneficiaries of the official developmentalist policy (Gonçalves 1992, Gonçalves et al. 1994).

A National Economic Development Bank (BNDE) resolution gave priority to financing huge projects with a daily pulp production capacity of 1,000 tonnes or more. Such lending arrangements, naturally, also favoured big plantations. Indeed, loans were to be made available for mill construction only in those cases in which half of the fibre required was already available (Gonçalves et al. 1994). In addition to investing public funds in the pulp industry, the state also played a central role in creating institutional and legal conditions necessary for plantation development and in spreading the idea that pulp projects would guarantee progress wherever they were set up (Miranda 1993a). In some cases state companies and cooperative ventures between the state and private capital have also been involved (see below).

The new pulp businesses were situated mainly in the east and north of the country. Such locations, since they made it possible for factories to be sited near large raw-materials sources, allowed the industry to take advantage of economies of scale and integration. They also enabled the industry to avoid the large-scale popular resistance to pollution which would have resulted if factories were sited near major urban centres. By building mills near the coast, finally, the industry was able to keep transport costs to export ports low (Gonçalves 1992).

An apparent success

Brazil first began making pulp and paper on a considerable scale in the 1950s, when annual pulp production reached 280,000 tonnes. During the 1960s, the figure rose to 780,000 tonnes, but it was in the 1970s that the sector really took root. In 1980, more than three million tonnes was produced, with short-fibre eucalyptus predominating over long-fibre conifer pulp. In the 1980s, when the industry's motor became export pulp production, total output rose to four million tonnes. By 1994, Brazil was supplying half the world's eucalyptus pulp (Gonçalves et al. 1994, ABECEL n.d.). Today the country is seen as the home of the outstanding examples of big, industrial fast-growing tree plantations, harbouring the second-largest area of such plantations in the world. Ballyhooed as the 'leading supplier of eucalyptus market pulp in the 1990s', it ships nearly two million tonnes abroad each year (*Know-How Wire* 2.1993, ABECEL n.d.) and boasts pulp-making costs among the lowest anywhere (Shell/WWF 1993).

This enormous expansion of the pulp sector, ironically, has taken place in a country with a per capita paper consumption which remains well below the world average. While yearly per capita paper consumption in the US is 332 kilogrammes and in Switzerland 216, Costa Rica 55 and Argentina 45, in Brazil it is only about 28 (*PPI* 7.1995). Most Brazilians have not benefitted from increased paper consumption; nor have many new jobs been created by the industry. Overall effects, environmental and other, have been negative.

This prompts a question: in what sense can Brazil's massive pulp and plantation development be considered a success? To try to answer this question, this chapter will examine the history, claims and records of five of the most frequently-praised of the country's pulp operations — Aracruz, Bahia Sul Celulose, Monte Dourado, Celulose Nipo-Brasileira (CENIBRA) and Riocell — which, among them, produce 90 per cent of Brazil's bleached eucalyptus pulp exports.

Aracruz

Aracruz is the world's largest producer of bleached eucalyptus kraft market pulp and operates the world's largest pulp mill. In 1978 it inaugurated its first 400,000-tonne-per-year plant (later enlarged to 500,000 tpy) and in 1991 the second, which produces a further 500,000 tonnes (Willums and Goluke 1992). Aracruz's production is aimed primarily for export, with Europe as the main destination, followed by the United States and Japan. The company accounts for more than half of Brazil's pulp exports, or more than 850,000 tonnes (Miranda 1993a, ABECEL n.d.). European customers include Jamont, Arjo Wiggins Appleton, Sappi, Inveresk, Tullis Russell, and UK Paper.

Aracruz's mills feed off its extensive eucalyptus plantations in the states of Espírito Santo and Bahia, where the company has achieved increasingly rapid growth rates through the selection and cloning of the fastest-growing species and the individual trees best suited to the needs of industrial processes (that is, those with the least bark and branches). Wood consumption per tonne of pulp produced thus dropped from 4.5 to 4.1 tonnes between 1985 and 1991 (Swann 1993). At the same time, average tree growth in the plantations has gone from 25 to 35 cubic metres per hectare per year (Willums and Goluke 1992).

Although ostensibly a highly successful private undertaking, Aracruz was in fact conceived and developed through direct and indirect state assistance. One of Aracruz Florestal's original 1967 partners was the ECOTEC consultancy company, which not only pressed the government into approving fiscal incentives for plantation forestry, but also advised the government of Espírito Santo state to promote the activity. ECOTEC also recruited entrepreneurs to back the business and chose the location of the plant. In 1974, the government approved a decree allowing an 80,000 hectare afforestation project to be set up in Espírito Santo. One of the signers was the then Minister of Mines and Energy, Antonio Dias Leite Jr. — an ECOTEC partner (Gonçalves et al. 1994).

Established in 1968, the plantations soon fell victim to disease. Aracruz responded by sinking money into research on seeds of various provenances, becoming a recognized authority on the subject from the 1980s onwards (Gonçalves et al. 1994). By the middle of the 1980s, meanwhile, the company had acquired so much land in Espírito Santo that the state government was forced by the public to come to a 'gentlemen's agreement' with the firm to prevent it from buying more. Aracruz's response was to invest in 80,000 to 100,000 hectares in neighbouring Bahia. A new Forest Development Programme, moreover, allowed the company and the state to promote eucalyptus planting on the land of third parties (CEPEDES/CDDH 1992, FASE 1993). As of 1994, Aracruz owned 1.86 per cent of Espírito Santo's territory, commanding in all 203,000 hectares there and in Bahia. Of this, 132,000 hectares had been planted with eucalyptus (IBASE 1994b, Aracruz 1996).

Aracruz Cellulose was set up in 1972 and Aracruz Florestal became its subsidiary. The first shareholders were a widely varied group of banks, oil firms, construction companies, and other businesses, none of which had any previous experience in pulp production. Among them was the Lorentzen Group (one of the few which is still a major shareholder in the company). To boost its expertise, Aracruz formed an association with the Swedish Billerud firm, which had some experience in eucalyptus pulpmaking and which took responsibility for marketing half of Aracruz's output in Europe (Goncalves et al. 1994).

In 1975, the state's BNDE agreed to contribute US$337 million, in the form of loans and share purchases, to the construction of Aracruz's first mill. The financial package, granted on exceptionally favourable terms, was the largest ever offered to a private firm. The state became the company's main shareholder, with a 40 per cent stake, followed by British American Tobacco with 30 per cent, and in all provided 55 per cent of the support needed to build the mill. Without this state support, the plant could never have been set up (Goncalves et al. 1994).

Following further BNDE loans to increase plant capacity, a privatization process got under way in 1985, ending in 1989 with the sale of the majority of state-held shares. Current main shareholders are Souza Cruz (a subsidiary of British American Tobacco), the Lorentzen Group, and the Safra Group, with 28 per cent each; the holdings of BNDE (now renamed BNDES) have been reduced to 12 per cent. According to the research findings of FASE and IBASE, two major Brazilian non-governmental organizations (NGOs), while privatization 'represented a good deal for those groups assuming control of the company', it was an 'absurd move' from the point of view of the state, which thereby abandoned its strategic position in a lucrative undertaking, handing the taxpayers' inheritance over to the private sector (Goncalves et al. 1994).

Aracruz, perhaps more than any other big pulp maker, has taken seriously the need to respond to pressure from the environmental movement — above all through publicity. The firm's directors played a leading role, through the Business Council for Sustainable Development, in preparing big business for the Rio Earth Summit in 1992. This resulted in a book, *Changing Course*, which was distributed to all government delegations before the conference. The company's environmental publicity has been widely disseminated, both by the company itself and through the Business Council for Sustainable Development, as well as by journalists and researchers. Aracruz has also supported the London-based non-governmental International Institute for Environment and Development in a large study on the 'sustainable paper cycle' (IIED 1995).

In its publicity, the company claims not only that it uses 'clean' industrial processes. It also asserts that it has planted eucalyptus only in areas where the natural forest was cleared before its arrival. The firm stresses, in addition, that it is making every effort to conserve remaining native stands. It states that 27 per cent of its land has original forest cover and that for every 2.4 hectares of eucalyptus one hectare of natural forest is preserved. To counteract forest defence campaigns, Aracruz highlights the fact that it has always used wood from its own plantations or has bought eucalyptus wood from farmers provided with free seedlings by the company (Willums and Goluke 1992). At the same time, it states that its plantations do not compete for land used for agriculture or other activities (*Know-How Wire* 2.1993). Perhaps no one has promoted this image better than Sargent and Bass in their book *Plantation Politics* (1992). After describing the panorama of desolation in Espírito Santo and Bahia, these authors go on to say that 'Aracruz Celulose S.A., with government support, took control of much degraded land within the tattered fragments of natural forest, and has established major eucalyptus plantations. In doing so, it has begun to improve the local environment and social conditions'. So successful has Aracruz publicity been, indeed, that in 1991 the company was officially singled out by the United Nations for its 'rigorous commitment' to sustainable development (IBASE 1993a).

This chorus of praise, however, has little factual foundation. Aracruz's tenure has had deleterious effects both on local people and their livelihoods and on the soil, water and forests of the region.

First, the enormous area bought by the company beginning in the 1960s was not, contrary to company claims, empty: thousands of indigenous people and subsistence farmers lived there. In order to overcome local resistance to the takeover, mainly from black communities and small agricultural producers who had recently migrated from other states, the company used a strategy in which physical violence was joined to symbolic violence. Land purchases were made through two people: a military officer and a local black leader, a combination which had an especially clear meaning given the dominance at that time of the military dictatorship (Miranda 1993b). Approximately 7,000 families are said to have been removed, through violence and coercion, from the areas Aracruz occupied, including several thousand people who received no compensation (FASE 1993, Valarelli 1992, *IAD* 7/8.1992). Crucial to these expulsions was the negligence or collaboration of local authorities, including the then governor of Espírito Santo state, Artur Gerardt, who later became president of Aracruz (FASE 1993).

In 1967, recalled José Luis Ramos, the head of Caieira Velha village, Aracruz, with the support of the state government,

> felled large areas of forest and planted eucalyptus across the region, including on our land. In a little time, this company destroyed around seven Tupiniquim villages, expelling us, and today we are surrounded by a sea of eucalyptus in three discontinuous areas. Now there is no forest or game left. The valley's rivers disappeared following the planting of the eucalyptus (Miranda 1993c).

Three principal options were left to the people expelled: emigrate to other rural areas; move to a life of underemployment in the *favelas* or shanty towns of cities; or work for the company, mainly on the plantations. Aracruz plantation work, however, has been described as near-slave labour, with conditions so dangerous and unhealthy that few workers can remain on the job more than ten years. Over 50 per cent of those who have worked on the plantations long-term suffer from serious work-related health problems as a result of their tasks. While they have the right to medical treatment, it is rare that they can ensure that they are given it, since the doctors are also Aracruz employees. The company tends to dismiss long-term plantation employees without compensation, replacing them with younger people (Inyaku 1993). The abundant reserve army of unemployed — whose numbers have been swelled by the crisis of another export monoculture crop, coffee, by the company's dispossession of local people, and by migrants attracted to the Aracruz 'development centre' — meanwhile helps keep salaries low.

Yet opportunities for even low-paid, health-endangering jobs have become limited as tree harvesting and other plantation work have become increasingly mechanized (IBASE 1994b). During the first months of 1993 alone, Aracruz dismissed 20 per cent of its employees, reducing the workforce from 7,000 to less than 6,000, with forestry jobs being most affected (Inyaku 1993). In 1993, the indigenous people who had lost their land to Aracruz started a new battle to get it back.

Aracruz attempts to divert attention from its record by advertising itself as having voluntarily built several recreation centres, schools, vocational training centres and health centres, at a cost of over US$15 million. What the company does not mention, however, is that as part of the plant expansion project, the BNDES both demanded that Aracruz reinforce social structure in the regions where it operated and put up 80 per cent of the money required (Gonçalves et al. 1994).

Aracruz's claim that it does not usurp agricultural land is contradicted by the way the firm has bought land in Bahia previously dedicated to small-scale cultivation of mamao, a local fruit. Although a company director suggested that mamao production was in decline, it had in fact provided a viable livelihood for local farmers (Gonçalves et al. 1994). The company's insistence that its activities are compatible with agriculture makes little sense given its propensity for buying precisely the best agricultural land on the grounds that it makes mechanization easier. Aracruz owns a full 15 per cent of the plains in Espírito Santo (FASE 1993).

On the environmental side, meanwhile, Aracruz has been accused of having felled and burned more than 50,000 hectares of forest during its first phase of tree-planting (FASE 1993, Miranda 1993c, *IAD* 7/8.1992). It has also been fined by IBAMA, Brazil's environmental protection agency, for planting eucalyptus in protected areas. Rogério Medeiros, National Coordinator for the Environment of the National Federation of Journalists, who has been monitoring Aracruz's activities since its beginnings, states that 156 streams have disappeared in the region during that period and that wells are drying up in a number of areas; even a river, the San Domingos, has stopped flowing. In the words of Joao Pedro Stedile of the Landless Workers Movement, '50,000 people in the area used to eat fish every day. Now they eat fish no more; some fishermen have stopped fishing because there are so few fish to catch' (*IAD* 7/8.1992).

The Tupiniquim Indians, meanwhile, found that scattered parcels of land grudgingly returned to them in 1983 after having been planted to eucalyptus did not return to their old level of agricultural productivity, their soils having become sandy. Indiscriminate use of the herbicides glyphosate and Goal and the ant killer Mirex was blamed for the poisoning of animals in the region (FASE 1993). (Since 1993, apparently, all major pulp companies have used Mirex-S, a less harmful piretroid, instead of Mirex, an organochlorine [Fanzeres 1995].)

According to the local NGOs, Centro de Estudos e Pesquisas para o Desenvolvimento do Extremo Sul da Bahia and Centro de Defesa dos Direitos Humanos, Aracruz has in addition been 'constantly fined' for industrial contamination. Local people have testified that

> near the factory it is nearly impossible to breathe. The majority of the people working in the factory probably have no nasal mucous left . . . pollution emitted by Aracruz reaches distances of more than 100 kilometres, depending on the wind (CEPEDES/ CDDH 1992).

The Pulp Industry Workers Union (SINTICEL) has accused Aracruz of incorrect management of both waterborne and airborne emissions (FASE 1992). Local activists claim that chemical releases into the Atlantic have poisoned both fish and vegetation (*IAD* 7/8.1992).

Moreover, Aracruz's newly professed concern with implementing 'clean' industrial processes cannot be expected to lead to consistent environmental results if it consists mainly in an interest in profit margins. When the company incorporated new technology for non-chlorine pulp bleaching in late 1993, it was simply responding to foreign consumer demand for chlorine-free pulp. A company directive stated that 'without the new technology, we would lose sales of 150,000 tonnes of pulp on the international market' (IBASE 1993b). It is for this reason that the firm runs three different types of industrial process among its four production lines. One uses chlorine gas, two others are elemental chlorine free (ECF) and only the newest one is totally chlorine free (TCF) (Fanzeres 1995).

Small wonder that in May 1992, on the eve of the Earth Summit, the Brazilian NGO Forum, together with representatives from other sectors of society, organised a public event denouncing the 'Aracruz farce' and criticizing a proposal to present the company's activities as a model of sustainable development. At the same time, Greenpeace symbolically closed Portocel, the firm's export port, as a sign of protest against the environmental and social damage the firm had inflicted during pulp production, including its destruction of native forest (FASE 1993, Bruno 1992). As one research team concluded, Aracruz

> assumes the image of protector of the environment, but its eucalyptus trees have dried streams, destroyed the local fauna, impoverished the soil, impeded the regrowth of native plant species, and drastically reduced the area available for cultivating basic foodstuffs (in a country where many people die of hunger).

This is not to mention land concentration and the expulsion of the rural population, which has contributed to increasing the urban population and the degradation of living conditions in the cities. Where is the sustainable development here, we might ask (Gonçalves et al. 1994)?

Bahia Sul Celulose

Bahia Sul Celulose was set up in 1984 by one of the main business conglomerates in Brazil, the Suzano Group, and the state-owned Companhia Vale do Rio Doce. Suzano, with interests in many areas, is owner of the Suzano Paper and Pulp Company, one of the main integrated paper-producing concerns in the country and involved also in wood, publishing, and graphics. Previously, Suzano Paper and Pulp had produced pulp only for its own paper mills; Bahia Sul was its first venture into export pulp.

Bahia Sul's shareholders include: Suzano (CSPC), which today holds 35 per cent of the shares; Companhia Vale do Rio Doce (CVRD), with 29 per cent; the National Economic and Social Development Bank (BNDES), with 26 per cent; and the International Finance Corporation (IFC) of the World Bank, with three per cent. Voting rights have always rested with Suzano and CVRD, who hold 55 and 45 per cent of the ordinary shares.

Total investment came to US$1.5 billion, with the company benefiting from ten years of low-interest BNDES loans worth $387 million, all of which had a three years' grace period. Bahia Sul also obtained $33 million from Citibank and $83 million from Manufacturers Hanover Trust. IFC involvement, by signalling World Bank support, made it easier to raise funds from other sources (Gonçalves et al. 1994).

Bahia Sul Celulose has the capacity to make 500,000 tonnes of bleached eucalyptus pulp and 250,000 tonnes of printing and writing paper per year. Some 45 per cent of its pulp is earmarked for the firm's own paper mills, another 45 per cent for export and the last 10 per cent for the domestic market (CEPEDES/CDDH 1992). Half the export pulp goes to Asia and another quarter each to North America and Europe. Three-fifths of the company's paper production is also shipped abroad, half of it to North America and equal parts of the rest to Europe and the Middle East (Gonçalves et al. 1994). The firm ranks fifth among Brazil's pulp exporters.

The decision to locate the company in Bahia was motivated largely by the existence there of extensive eucalyptus plantations. Suzano, a pioneer in eucalyptus pulping, had exhausted its expansion possibilities in São Paulo state and wanted more pulp. CVRD, which was already embarking on joint ventures in the pulp sector, and which held 100,000 hectares of eucalyptus in Bahia originally planted for charcoal for the steel industry, made an ideal partner (Higgs 1992a, 1992c).

The CVRD plantations were the result of a government incentive policy promoting tree-planting in the central southern region of Brazil. At the end of the 1970s, land prices rose in those regions and plantation firms moved into southern Bahia, where large areas of cheap or deforested land were becoming available in the wake of highway construction and the associated unsustainable logging. Their goal was less to set up productive enterprises than simply to get hold of the subsidies that were being poured into the plantation sector. In 1982, the state of Bahia became the second largest afforested region in Brazil (Miranda 1992). All the conditions for a large export pulp mill were then available: abundant raw materials, cheap land, a cheap workforce, transport infrastructure and a nearby port in the city of Vitória.

Today Bahia Sul Celulose owns 114,000 hectares of land, divided between 68,000 of eucalyptus, 34,000 of native forest and 12,000 occupied by infrastructure works. The company plants an additional 1,000 hectares annually and since 1992 has also promoted tree-growing by small farmers on their own land in order to broaden the resource base with a view to eventual expansion of the plant (Gonçalves et al. 1994, Higgs 1992c).

Like Aracruz, Bahia Sul Celulose presents itself as a defender of nature and blames the poor for ecological degradation. The hunting practiced by local people to supplement their poor diet, for example, is treated in the company's literature as an 'ecological crime' which the firm is heroically striving to curb. An industry journalist reported approvingly in 1992 that 'environmental considerations were important both in the location of [Bahia Sul's] plant and in its construction':

> Apart from Bahia Sul's conservation of 30 per cent of its forest land for native species and its planting of 20,000 such trees annually, extensive studies were produced prior to construction to assure the aquatic environment was safe. Features of the mill include oxygen delignification, primary and secondary effluent treatment and biomass energy production (Higgs 1992).

Such statements are misleading. They ignore the fact that, before the region was opened up to 'development', local people had been using natural resources in a far more responsible manner than the Bahia Sul is using them today (Miranda 1992). They are also contradicted by the fact that the firm used only 7.2 per cent of its total mill construction costs on contamination control equipment (Gonçalves et al. 1994), as opposed to the approximately 20 per cent figure common in industrialized countries (CEPEDES/CDDH 1992).

Also unmentioned are other socioeconomic dangers the Bahia Sul installations pose for the region as a whole, such as those related to dependence on a single monoculture. In the case of cacao cultivation in Bahia, this dependence led to repeated crises attributable to climatic irregularities, pests, and falling international commodity prices over which local producers had little control. Overdependence on monocultures of eucalyptus may bring similar problems. Reductions in the state's agricultural area have already forced it to import more expensive food from elsewhere, with negative repercussions on the quality of life of the worse-off (CEPEDES/CDDH 1992, Miranda 1992).

To make way for Bahia Sul, in addition, the families from nearly 8,000 small properties in the region were driven from their land. Evictees who became forestry employees are now by and large worse off than small producers who managed to keep their properties (Gonçalves et al. 1994).

Nor has Bahia Sul proved to be an efficient creator of new jobs. Each of the existing 5,500 jobs at the firm has required an investment of between US$226,000 and $338,000 — much of the capital deriving ultimately from public funds. According to a study conducted by local NGOs (CEPEDES/CDDH 1992), an alternative development project (based on diversified agriculture) with a similar level of investment could generate 150,000 jobs at a cost of a mere $8,300 each. In the forestry sector, the company employs only one person per 45 hectares, while agricultural activities give work to at least 18 per hectare; mamao production requires 1.5 workers per hectare (CDDH 1993). In spite of company propaganda, moreover, many of the jobs which have been provided have not gone to local people, who are often not sufficiently qualified (Miranda 1992).

Living and working conditions have improved little. The large numbers of people attracted to the Bahia Sul site by the company's advertisements for construction workers pushed up housing rents. For labourers who lodged alone, living conditions were dreadful. Rooms measuring three by four metres lodged six or more workers, with one toilet per 32 men. Family problems were exacerbated by extended separations. Some 5,000 workers lived in such conditions for three years, with the population around the construction site occasionally surging as high as 13,000. The resulting strikes delayed completion of the plant. Inmigration also contributed to the collapse of health, education, drainage, housing and security infrastructure. New shanty towns sprang up, and poverty, violence, crime, disease and prostitution increased (CEPEDES/CDDH 1992, CDDH 1993, Gonçalves et al. 1994).

Trade unionists have meanwhile pointed to high unemployment and underemployment and at poor working conditions in the forestry sector

(Miranda 1992). The transport system used for both the construction works and forestry areas has been denounced as degrading and dangerous. Workers have been treated as livestock and transported in the back of lorries, resulting in frequent injuries. Bahia Sul has also damaged local fishing and tourism livelihoods (CEPEDES/CDDH 1992).

A megaproject of the Bahia Sul variety implies enormous concentrations of capital, land and political power. As it generates little employment in relation to the level of investment, little of this capital and power ever finds its way down to society's deprived. As a result, large firms such as Bahia Sul come to wield disproportionate political clout and the states in which they are located run the risk of suffering from a modern 'feudal' syndrome in which large holders of land and capital become 'lords' who attempt to dictate policy. Before Bahia Sul even started up, official complaints were being aired about the use of company funds in election campaigns (CEPEDES/CDDH 1992). Years later, an advance payment of taxes by Bahia Sul was used by the Bahia government to surface the roads most used by the company's lorries (Miranda 1992c). In 1989, in a parallel with events in Espírito Santo, the state government of Bahia was belatedly forced, in reaction to the company's buildup of 'feudal' power, to demand that the firm suspend acquisition of new lands.

The environmental problems caused by Bahia Sul Celulose are also significant. In southern Bahia, where the company is located, only some 60,000 hectares survive of a forest which, in 1930, covered a million and a half hectares. These remnants are part of the Mata Atlantica, an endangered forest at least as rich in biodiversity as the Amazon rainforests (CEPEDES/CDDH 1992). The eucalyptus used by Bahia Sul threatens the survival of this forest mainly because it takes over crucial areas which would otherwise have regenerated into native forest. Unless these areas are allowed to regrow, the strict conservation of a few tiny areas of untouched Mata Atlantica forest — which, for public relations purposes, Bahia Sul takes great care to say it supports — has little meaning.

Bahia Sul's eucalyptus plantations are also affecting both the quantity and the quality of local water. Inhabitants of the town of Veracruz have declared that, as the result of a nearby eucalyptus plantation, several wells have dried up and they have had to dig down a further three metres in order to find water. Insecticides, fertilizers, and herbicides used on the plantations, in addition, contaminate watercourses, taking a toll on aquatic fauna. The inhabitants of the fishing community of Caravelas have taken legal action to demand an investigation into a recent reduction in crabs and other species, which has been linked to the use of agrochemicals (CEPEDES/CDDH 1992).

Local residents fear that water contamination from Bahia Sul's industrial plant will only add to the problems.

CENIBRA

The first eucalyptus plantations were established in Minas Gerais in the 1940s by the Belgo-Mineira Company, for charcoal. They expanded enormously from the late 1960s onwards (particularly in the 1970s), due largely to state subsidies. Plantations were set up by Florestas Rio Doce (created by the state and the CVRD in 1968) and the steel company Acos Especiais Itabira (ACESITA), as well as independent tree growers, the aim being to supply pulp, iron and steel and cement industries with raw materials or energy. Minas Gerais soon boasted the biggest eucalyptus plantation area in Brazil (M. T. Gonçalves 1995).

By 1970, Japan was CVRD's biggest customer. In 1972, CVRD signed an agreement with the Japan–Brazil Pulp Resources Development Co. (JBP) to set up a jointly held company called Celulose Nipo-Brasileira, or CENIBRA, to produce eucalyptus kraft pulp for export, half of which would go to Japan. CVRD held 51.5 per cent of the shares (JATAN 1993). JBP, the minority shareholder, was in turn over 36 per cent owned by Japan's bilateral foreign 'aid' agency, the Overseas Economic Cooperation Fund. Other JBP shareholders included Oji Paper (nearly ten per cent); Jujo Paper, Honshu Paper and Itochu (over eight per cent each); Tokai Pulp, Daishowa Paper, Hokuetsu Seichi Paper Mills, Mitsubishi Paper Mills, Daio Paper, and Sanyo Kokusaku Pulp (over 3.5 per cent each); and, with smaller stakes, Kanzaki Paper, Daiko Paper, Chuetsu Pulp, Nippon Kakou Sheisi, Abegawa, Kishu, Sanko, Juban and Mishima (M. T. Gonçalves 1995).

In 1974, when CENIBRA began operating, its pulp mill was supplied with wood from Florestas Rio Doce, the CVRD subsidiary. At the same time, another firm called Florestas Nipo-Brasil, or FLONIBRA, was created, with the aim of acquiring an additional 400,000 hectares of land for pulp plantations in Bahia and Espírito Santo. In 1984, the 155,000 hectares which had actually been acquired — including 88,000 planted with eucalyptus and 48,000 given over to 'conservation areas' — were passed to CENIBRA Florestal, the firm now responsible for supplying the CENIBRA mill with pulpwood and energy (Gonçalves et al. 1994). Some 4,600 people are employed on the CENIBRA Florestal lands. Among other tasks, they plant 6,000 hectares of eucalyptus per year (JATAN 1993).

CENIBRA's pulp mill, which employs around 1,300 people, increased production from 116,000 tonnes in 1978 to 376,000 in 1993, and plans to double that figure have already been approved (JATAN 1993, Gonçalves 1995). Fully 80 per cent of the company's pulp is exported

(50 per cent to Japan and 30 per cent to Europe and the US), making the firm Brazil's second-ranked pulp exporter after Aracruz (JATAN 1993, Miranda 1993a).

The firm's inland mill in Minas Gerais is able to compete with coastal export pulp operations such as Aracruz only because it has the use of a high-quality railway belonging to CVRD. When the plant came on line, the railway was improved and adapted to pulp transport (Goncalves et al. 1994). The company is also able to keep costs down through being co-owner, together with Aracruz, of Portocel, a port especially adapted for pulp handling (Knight 1991b). In 1975, moreover, CENIBRA received US$128 million in state financing. This was supplemented by another $14 million in 1976, $16 million in 1981 (to reduce energy consumption), $6.6 million in 1982, and $15 million in 1985 (Gonçalves et al. 1994).

CENIBRA's claim to green credentials comes in the form of a five-year strategic plan developed in 1991 to assure 'sustainable development' (Higgs 1992d). As a result of this plan, 240 hectares of pine and eucalyptus surrounding the plant were to be enriched with local species in order to encourage the return of other indigenous species. In addition, some 1,000 hectares of native forest were to be preserved, while company land on the banks of the Rio Doce was to be reforested with native species. At the firm's mill, meanwhile, the plan called for company-monitored measures to control effluents, as well as a start to production of chlorine-free pulp.

Several aspects of this plan call for comment. First, the planting of indigenous species and the preservation of a few areas of native forest (the latter, far from being a voluntary measure, is a legal obligation) are essentially cosmetic measures to pacify local people increasingly concerned by the advance of eucalyptus, as well as environmental groups in the North. (They may also limit, to a small extent, the risk of pests.) Second, effluent-treatment systems were installed only in 1988 — 11 years after the mill came on line — and only after heavy pressure from local people (M T. Gonçalves 1995). Their installation thus hardly demonstrates trustworthiness on environmental matters. Similarly, chlorine-free bleaching techniques began to be used only as a result of demand from the European Community, and are used exclusively with pulp destined for the European market (JATAN 1993).

As with Aracruz's and Bahia Sul Celulose's 'greenwashing' efforts, moreover, CENIBRA's 'sustainable development' plan artfully fails to mention a number of destructive realities about the company's operations. One such reality is the way CENIBRA's projects and those of other companies such as CVRD, ACESITA and Belgo-Mineira concentrate vast swathes of land in a few hands in a process often marked by violence. According to JATAN (1993), FLONIBRA 'used whatever methods were available to acquire land for its plantations':

> sometimes it purchased land at above market prices; at other times residents were chased away with violence. It also often resorted to deception; for instance it would first move a FLONIBRA insider onto land adjacent to the farmer's land and set up a local conflict; then a third party would be sent in who would act as a 'mediator' between the two and offer to purchase the farmer's land.

FLONIBRA is also reported to have felled secondary forest used by indigenous people, forcing emigration to nearby towns (JATAN 1992).

As elsewhere in Brazil, the concentration of land with good soils in a few hands has undermined subsistence agriculture and led to increased urban migration and the weakening of autonomy and local social ties (Guerra 1992, M. T. Goncalves 1995). Wage labour, meanwhile, has proved an inadequate substitute for small-scale farming in both economic and cultural terms. As one worker said, 'the salary that we earn does not go very far . . . it is only enough to survive on' (M. T. Goncalves 1995). As a result of large plantations taking over agricultural areas, Minas Gerais has had to import food from other regions. As the holdings of plantation firms accumulate, moreover (ACESITA, for example, has 250,000 hectares, Belgo-Mineira 100,000 and CENIBRA 155,000), the dependence of local towns on a few businesses grows, and the influence of such companies on decision-making processes increases. One result is yet more industrial projects and tree plantations (Guerra 1992, JATAN 1993).

CENIBRA's relationship with its workers, meanwhile, has been authoritarian and exploitative. While, according to one interviewed worker, wages were initially good and there was a strong trade union, salaries declined considerably after CENIBRA stepped in to manipulate union elections through pressure and fraud. The company also dismissed unionized workers and started contracting out both industrial and forestry tasks to other firms. This move, aimed at reducing costs, resulted in even lower wage levels and a smaller worker population, making unionization even more difficult (M. T. Gonçalves 1995). Working conditions are often inadequate, with many plantation chainsaw operators suffering from nervous disorders and other health problems (JATAN 1993).

CENIBRA is also moving into contract farming through the Fazendeiro Florestal programme, operated in conjunction with a state agency, through which seedlings, fertilizers and ant poison are provided to individual farmers if they plant eucalyptus on their own land. This allows the company to augment its forestry base in a way other than through simply acquiring land. Farmers are contractually obligated to sell wood to CENIBRA at the 'market price', which is set by the company itself (M. T. Gonçalves 1995).

Environmental impacts of CENIBRA plantations include destruction of native forests and loss of biodiversity (Guerra 1992, JATAN 1993, M. T. Goncalves 1995). Fertile farmland has also been taken over. Not only does CENIBRA use non-'degraded' land; it also plants, contrary to stated policy, on pronounced slopes (Guerra 1992).

Soil ecology has also been affected. Because the thick layer of plant material associated with plantations cannot be mineralized rapidly by microorganisms, organic acids are formed and calcium, potassium and magnesium ions are replaced by hydrogen ions in the upper layer of the soil. This implies a lower pH, which affects the availability of nutrients to plants. The long term productive capacity of the soil is endangered and it is not known how much longer the same land can keep on producing eucalyptus wood (Guerra 1992). Tree bark, which, according to FAO, contains the best part of the nutrients taken from the soil by the tree, is removed from the site and used in the pulp mill for energy generation, further reducing soil fertility (JATAN 1993).

Local people have observed, moreover, that the rate of replacement of underground water tables has slowed. This is attributable to the increased surface runoff associated with plantations and with eucalyptus's high water consumption. Finally, the use of pesticides such as aldrin and Mirex has led to contamination of soils and watercourses, with proven high fish mortality (Guerra 1992). The company has even begun to spray herbicides from the air, endangering local agriculturalists (M. T. Gonçalves 1995).

Environmental impacts at CENIBRA's pulp mill have been equally serious. Complaints about water pollution have been continuous since 1977, the year the plant started up. Yet only at the end of 1986 did the company formally commit itself to installing an industrial and sanitary effluent treatment system, giving itself a 30-month deadline. Since 1990, CENIBRA has submitted monthly reports to the appropriate state agency, but without making them public (M. T. Gonçalves 1995). Although scientific data on air pollution around the mill is unavailable, JATAN members visiting the plant in 1992 remarked that 'we couldn't avoid noticing the horrible smell present in all of the plant and we felt sorry for the people who had to work there' (JATAN 1993).

Jari and Companhia Florestal Monte Dourado

In 1967, United States shipping magnate Daniel Ludwig paid the Brazilian military dictatorship US$3 million for 1.6 million hectares of tropical forest on the Jari River near the frontier with French Guiana.

A year later he began planting vast homogenous masses of fast-growing trees for pulp. Ludwig's pulp mill, constructed in Japan and taken by sea to the chosen site, was finally installed in 1979.

Ludwig's project was similar to others in Brazil, except for the species chosen. This was *Gmelina arborea*, a tree indigenous to Indonesia, and was a failure. In addition to the problems normally experienced with eucalyptus and pine (leaf-cutting *Atta* and *Acromyrmex* ants), the gmelina was afflicted by the fungus *Ceratosystis fimbriata*, which not only delayed tree growth, but also reduced wood quality (Higgs 1993b, Shell/WWF 1993). These difficulties caused the company great losses, as by 1978 it had already planted 64,000 hectares of gmelina (Shell/WWF 1993). From then on it began to plant a pine (*Pinus caribaea* var. *hondurensis*), which had already been tried in the area.

In 1982, when Brazil was moving towards democracy and Ludwig could no longer count on support from the military (Sargent and Bass 1992), he decided to sell Jari to a consortium of 23 Brazilian companies, led by the CAEMI group, who acquired 40 per cent of the shares (IBASE 1993a). The BNDES supported the operation through a US$180 million loan (Shell/WWF 1993). One of the main new shareholders was AMCEL (Amapá Florestal e Celulose S.A.), owner of some 90,000 hectares of pine plantations near the plant (Knight 1991b).

The latter were of great importance to the company. As its own existing plantations were so unproductive, it had no other raw material sources to rely on until its new pine, and then eucalyptus, plantations matured (Knight 1991b). At present the firm owns some 90,000 hectares of plantations, which are mainly eucalyptus and pine, with a 2,000 hectare remnant of gmelina. Eucalyptus now dominates, 60 per cent from selected clones and 40 per cent from seed. More trees are being planted at a rate of 13,500 hectares per annum (Higgs 1993b).

Jari's mill — operated by Companhia Florestal Monte Dourado — produces 300,000 tonnes of pulp annually, 55 per cent of which is long fibre and 45 per cent short. Some 65 per cent goes to Europe, 14 per cent to the US, and another one per cent to increasingly important markets like Japan, Indonesia, Korea and China (Shell/WWF, 1993).

Jari has not lagged far behind its fellow firms in attempting to 'greenwash' its activities. As journalist Richard Higgs writes, 'Jari is very proud of its almost surgical blending of plantation species in among the predominant and thriving natural forest' — a practice which helps 'to prevent the spread of pests and disease', something of which the company has bitter experience — and maintains reserve areas and carries out research into native species which may be of economic use. Like other pulp firms, Jari has also recently displayed much concern about chlorine pollution, although this concern had lain dormant for 14 years before 1992, when European consumers began calling for non-chlorine-bleached pulp (Higgs 1993b). Jari's own public relations efforts have received a boost from Paulo de T. Alvim, a leading agricultural planner, who has claimed that the Jari plantations reduce global warming because they grow, whereas the tropical forest that had occupied the site previously had been in equilibrium with the atmosphere, neither absorbing nor emitting carbon (Fearnside 1993).

Such claims conceal a good deal more than they reveal. For example, although the company is legally obliged to preserve 50 per cent of the native forests it controls, it has already felled around a third of them in order to supplement shortfalls in gmelina with species such as *Jacaranda copaia*, which in 1982 made up around 20 per cent of its wood supply. Some 1,200 hectares of the forests Jari claims to be protecting are being felled each year (Shell/WWF 1993), mostly for energy generation and for the expansion of its plantations (Fanzeres 1995). In 1992, Brazil's official environmental conservation organization, IBAMA, rejected the company's request to cut 5,000 hectares of dense forest which it wanted to replace with plantations (IBASE 1993a). Felling such native forests, of course, has resulted not only in loss of native trees but also in loss of habitat for many other species.

Other environmental problems centre on the project's continuing reliance on monoculture. Forestry experts never tire of pointing out Ludwig's serious mistake in choosing gmelina as his plantation species (Sargent and Bass 1992, Shell/WWF 1993). This 'error', however, is merely one instance of a more general problem which foresters seem far less eager to acknowledge, and which current Jari plantings also exemplify: namely, that in a large-scale monospecific plantation of *any* fast-growing species, a fungus, virus, insect, or other animal pest may well be able to decimate the entire plantation in a short time.

Pest infestation, moreover, was only one of the environmental problems afflicting Jari. When the gmelina felling rotation was shortened to three to four years to avoid fungus attacks, nutrients began to be extracted from the soil at a higher rate. It is estimated that most of the potassium and phosphorus will have disappeared from the estate by the end of the 21st century (Shell/WWF 1993). The use of heavy forestry machinery, meanwhile, has caused soil compaction and erosion (Sargent and Bass 1992).

Nor has the company shown itself to be particularly concerned about water and aquatic life. Effluents from the plant are dumped directly into the Jari river, resulting in fish kills downriver. Fertilizers and agrochemicals used by the company also contaminate local watercourses (Shell/WWF 1993). Jari's 'environmental answer' to its energy needs — a hydroelectric dam on the Jari river (Knight 1991b,

Higgs 1993b) — would destroy one of the most beautiful and historic sites of Amapá state, Cachoeira Santo Antonio (Fanzeres 1995).

The claim that Jari's plantations help alleviate global warming, finally, has been convincingly refuted by scientist Philip M. Fearnside (1993), who points out that 'the much greater standing biomass of the forest [replaced by Jari] as compared to the plantation means that the effect of Jari is emission rather than removal of atmospheric CO_2'.

Jari's public relations exercises also conceal a pattern of widespread social abuse. When it was first set up, the company had to invest in social infrastructure to attract the huge workforce it needed. Some 3,000 housing units were built, as well as four schools, a 1,100-bed hospital, clinics, supermarkets, a radio station and 11,000 kilometres of roads (Higgs 1993b). Despite these investments, however, 'work-crew contractors were notorious for their treatment of workers brought in from the poor north-eastern states'. Staff turnover was constant at all levels, reaching rates of 200–300 per cent per year (Sargent and Bass 1992). After the company changed hands, new social problems emerged. Between 1988 and 1993 the worker numbers fell from 8,000 to 4,500. Many forestry labourers were replaced with machines. Many migrants were thus left unemployed in a region with few other potential employers. Meanwhile, the company began to pass responsibility for hospital, school and restaurant management onto local and federal authorities (Higgs 1993b): having attracted a large number of workers and their families to work on the project in its initial stages, it wound up shunting the long-term costs of their welfare onto the state. According to a local journalist, 'the legacy of the Jari project has been a shanty town in the middle of the jungle' (M. A. Goncalves 1995).

Riocell

Riocell, in the far southern state of Rio Grande do Sul, was created in 1975. The firm is owned by KIV, made up of Klabin, Votorantim, and the Iochpe finance company. Klabin is the world's 52nd-largest pulp and paper firm in sales. Votorantim, the country's biggest private conglomerate, took over the powerful Simao company in 1992. Holding interests in cement, aluminium and other metals, it has recently thrown itself into pulp and paper as well, becoming in 1994 the 65th biggest firm in the sector in the world (Higgs 1992b, Marcus 1993a, *PPI* 9.1995).

In 1975, Riocell bought its pulp mill from the Norwegian firm Borregaard, which had closed it earlier in the year as a result of public protests against the pollution it had caused since its construction in 1972. Apart from this factory, which is located in the city of Guaíba on the right bank of the Guaíba river 15 kilometres from the centre of the large city of Porto Alegre, the company owns 71,693 hectares of land, 53,000 of which are plantations (Schinke 1992b, Marcus 1993a). Riocell produces some 300,000 tonnes of bleached eucalyptus pulp each year and is the main pulp producer for the Brazilian rayon industry. The firm shares third place in Brazilian bleached eucalyptus pulp production and export with Monte Dourado (Jari), and is in the process of expanding its capacity to 700,000 tonnes per year. It has its own terminal in the port of Porto Alegre (Higgs 1992b, Knight 1991b).

Riocell has recently responded to domestic and foreign environment-al pressures with various 'green' claims. Klabin invites visitors to view company operations 'so they can see for themselves that we are not cutting down the rainforest'. Alfredo Lobl, a Klabin director, has stated that of the 330,000 hectares owned by the company, some 100,000 are preserved as native forests. 'We support environmental education programmes for 18,000 school children', Lobl adds (Marcus 1993a). Fellow director Celso Foelkel has insisted that 'rather than plant huge tracts of monocultural eucalyptus, the company has tried to integrate its growth as far as possible into the countryside ... nobody can say Riocell has a green desert' (Higgs 1992b). Riocell, like other Brazilian pulpmakers, has also begun to produce non-chlorine-bleached pulp.

These 'environmentalist' moves, however, simply make a virtue of commercial necessity, hardly reflecting a thoroughgoing commitment to change. It is the difficulty of finding large contiguous tracts of land for planting in Rio Grande do Sul, for example, and not a policy to 'fit into the area', that has forced Riocell into a pattern of dispersed holdings across the region. Similarly, it is pressure from the European market, and not social responsibility, that is pushing the firm into non-chlorine bleached pulp. Explains director Alfred Freund: 'We decided to go this way [elemental chlorine-free pulp production] because we're market oriented. Europe's our main market and Germany's important to us' (Higgs 1992b). To discover where Riocell's priorities really lie, it is necessary to examine other aspects of its record.

When the firm started up in 1975, it claimed it had put US$100 million into an effluent treatment system. Yet according to environ-mentalists, the company installed a sludge treatment unit only in 1987 (AGAPAN 1992), and in 1992, it was still being accused of dumping over 60 tonnes of organochloride compounds yearly into the Guaiba river, which provides drinking water for Porto Alegre (Schinke 1992b).

In 1992, the company, opposed strongly by environmentalists and others in its attempt to secure official approval to double its production capacity, tried to bribe a local government official who was opposing the expansion. Failing, it went on to tell the state governor that if he did not approve the project, the company would move to another state. The governor finally passed the project when the state government received a US$170 million loan from the Inter-American Development Bank for the decontamination of the Guaiba river basin (Schinke 1992a). Riocell itself had financed studies at the local university which showed that fish bred in water contaminated with effluent from the mill had developed serious genetic abnormalities (AGAPAN 1992).

Other planned projects

Among the other plantation and pulp projects under development in Brazil are the following:

•   CELMAR, planning a eucalyptus pulpmaking capacity of 420,000 tonnes a year for Maranhao, is a partnership of the Risipar group (an association of Votorantim and Ripasa), with 55 per cent of the capital; CVRD, with 30 per cent; and the Japanese Nissho Iwai Corporation, with 15. Some 80 per cent of the firm's output will go abroad, and 60 per cent of the funding will come from the BNDES, the World Bank and Japanese companies (Matías 1994).

•   FLORAR, sited near Maranhao's Carajas railway, is partnered with Aracruz Celulose, CVRD, and various Nordic firms. The company's pulp mill will have a n annual capacity of 500,000 tonnes of bleached eucalyptus pulp, 80 per cent of which is to be exported (Matías 1994).

•   CELPAV, a São Paulo plant, was acquired half-built from the BNDES in 1988 by the Votorantim group. The mill has a production capacity of 280,000 tonnes of bleached eucalyptus kraft pulp, half of which is slated for export. The company draws its raw materials from the plantations of the Votorantim group itself, which amount to some 45,000 hectares (Marcus 1993b).

•   Champion Celulose, a subsidiary of the North American giant Champion, which has had a presence in Brazil since 1959, is now installing a new pulp and paper mill in Amapá. Some 200,000 hectares of land are to be acquired, half to be planted with eucalyptus (M. A. Goncalves 1995). By 1993 the firm had already planted 25,000 of the 50,000 hectares of eucalyptus destined to feed the plant (Knight 1991b). According to local sources, these plantations are replacing a rare Amazonian instance of *cerrado*, the native scrubland of central Brazil. Characteristic problems associated with land takeovers and expulsion of small farmers are reportedly occurring (Fanzeres 1995).

- Veracruz Florestal, created in 1992, is a subsidiary of the powerful Odebrecht S.A. holding, which is involved in civil construction, petrochemicals, industrial assembly, aviation, naval construction, and agrobusiness (Gonçalves et al. 1994). Based in Bahia, it aims to produce some 600,000 tonnes of bleached pulp per year, with a raw material base of 62,000 hectares of eucalyptus monoculture. The first 41,000 hectares were acquired through an exchange of lands between Veracruz and CVRD. Veracruz has been implicated in deforestation since it was founded (Faillace and Miranda 1993).

Conclusions

The forests and rainforests of Brazil, with their enormous biodiversity and valuable woods, are rapidly disappearing. Yet the resources invested in studying how to manage them sustainably are ridiculously small. Instead of exploring the immense diversity of environmental and socioeconomic realities in Brazil, experts are channeled into studying and promoting uniformity in large-scale cultivation of coffee, cacao, sugar, soya — and fast-growing tree monocrops.

In addition, the state is pouring billions of dollars into developing the export pulp sector and its enormous industrial plantations. These plantations, which are often held up as a triumph of private-sector development, would not have been possible without massive direct and indirect state intervention in the form of fiscal incentives, soft loans, support from BNDES and similar agencies, and the involvement of the state's Companhia Vale do Rio Doce. The coercive power of the state has also been extremely useful in transferring the land of indigenous peoples and small farmers who have no title to their property into the hands of large forestry concerns. Finally, the 1964–84 military dictatorship's developmentalist vision of 'progress' filling 'empty spaces' has been critical in enlisting popular support for large pulp and plantation projects in the 'backward' regions of the country. It is no coincidence that the period of greatest growth in Brazil's forestry sector was precisely this 20-year period.

The 'progress' associated with plantations, however, has benefited only a minority. Concentration of land and power, migration, social disruption, a worsening of the quality of life, and long-term environmental degradation have been the results for Brazil as a whole. Environmental arguments adopted by pulp and plantation businesses cannot entirely hide this reality, though they do succeed in confusing public opinion both nationally and internationally.

Chapter 8

Chile: A Plantation Model Imposed by a Dictatorship

Chile's forestry development is being publicized as a successful application of neoliberal policies (Lara 1992, Messner 1993). Over 1.5 million hectares of trees have been planted in the country, the starting point for an impressive increase in timber exports, which have diversified to include over 400 different products with markets in 80 countries (*PPI* 11.1993). Chile currently holds the largest planted area of *Pinus radiata* in the world (Lara 1992), approximately 1.2 million hectares (Pandey 1992). However, Chile's forestry approach has resulted in a reduction of quality of life in the plantation zones, and has been an important cause of environmental degradation. Despite its 'neoliberal' label, the Chilean model has been based on direct and permanent state participation in the creation of wood resources and industrial infrastructure as well as in the establishment of 'rules of the game' favourable to the interests of the most powerful economic groups, both national and foreign.

The native forests

A bit of history

When the Spanish invaders arrived, most of the current territory of Chile (especially the South) was covered by magnificent forests, inhabited by indigenous communities using them in a sustainable manner. Like all colonizers, the recent arrivals aimed to exploit local resources and dominate local populations. When the indigenous peoples resisted, the colonizers resorted to violence. The war was long and hard. As indigenous peoples used forests as refuges to defend themselves against attacks, the colonizers deliberately burned large areas of tree cover with the sole aim of removing the native population's hiding place (CODEFF 1992, Cruz and Rivera 1983).

Once control had been established over local peoples, the forests were further degraded for commercial gain, a process which continues today. 'Civilizing' the country, moreover, required the felling of large areas to provide land for pastures and agriculture. In the region of Malleco, Arauco and Cautín, 300,000 hectares of forest were cleared in only ten years for grasslands and crops. The main crop was wheat, whose cultivation led to serious erosion within a short period. In other areas, such as Puerto Montt–Puerto Varas, enormous stretches of forest were burned for agricultural colonization projects which were never successful (CODEFF 1992).

The new agents of destruction

In the last few years, two new agents of forest destruction have come to the fore: pine (and more recently eucalyptus) plantations, and the export of wood chips for paper pulp production.

One of the main arguments wielded by promoters of tree crops is that wood from plantations, by meeting part of market demand, will make it less necessary to fell trees in native forests. In the case of Chile, this argument has proved false. In fact, pine plantations have become an active factor in the *degradation* of native forests.

In 1965, the existing 200,000 hectares of pines were indeed providing a substitute for native industrial raw materials (Cruz and Rivera 1983, Leyton 1986). However, in 1974, a new forestry policy gave rise to a process of substituting pine plantations for native forests. By 1983 it was being reported that 'the destruction of the Chilean native forest [and its replacement] with plantations of radiata pine constitutes one of the most serious and pressing problems of mismanagement and conservation of natural resources in Chile' (CODEFF 1983). In 1992, it was estimated that 6,195 hectares of forests were being replaced each year by plantations (PAF 1992).

From 1986, exploitation of native forests was accelerated in order to produce wood chips for export (mainly to Japan) for paper pulp production. According to estimates made by the non-governmental organization CODEFF in 1992, exports in 1990 alone used some 19,000 hectares of forest land, for the most part destroying its productive potential, with many native forests being replaced with eucalyptus plantations. A study recently undertaken by Chile's Central Bank suggests that within 25–30 years, at most half of the present estimated seven million hectares of native woods will remain for productive use (*The Economist* 3.2.1996).

Native forest-based wood chip exports increased from 13,900 tonnes in 1986 to 1,702,900 tonnes in 1991, constituting in this year 55 per cent of total wood chip exports. The other 45 per cent was based on plantations of eucayptus (30 per cent) and pine (15 per cent). Eucalyptus wood chip exports have had an impressive growth, from nil in 1987, to 210,000 tonnes in 1988, 430,000 in 1989, 575,000 in

1990 and 920,000 in 1991 (CODEFF 1992). According to FAO (1994), Chile's total exports of chips, particles and wood residues reached 3,796,000 tonnes in 1993. Since 95 per cent of these go to Japan, it is perhaps not surprising that Mitsubishi — which boasts a wholly-owned subsidiary called Astillas Exportaciones Lta. which produces chips from old-growth forests as well as from joint plantation ventures — is the largest wood-chip exporter.

Genesis of the current model

The origin of pine monocultures

Although it possesses extensive forests made up of valuable species, Chile has based its forestry development in large part on the monoculture of *Pinus radiata*, a pine from the United States. The introduction of this tree to the country dates from early this century when it was tested for its possible use for coal mine pit props in southern Chile. The wood was not suitable for this purpose, but the fact that the tree could grow rapidly led to the establishment of large plantations in the region from 1930 onwards.

This process was initiated principally by social security organizations which planted pine extensively on arable and pasture land of low productivity. Following suit were financial societies linked to forestry businesses, who sold middle-class investors small individual parcels of one hectare in extensive plantations. Timber industries seeking a secure source of raw materials then became directly involved.

Between 1940 and 1959, 10,000 hectares per year were planted. This rate diminished to 6,000 around 1964 as a result of the paper industry's policy of pushing wood prices down, which discouraged non-integrated wood producers (Leyton 1986).

The initial model

Since 1965, the Chilean state has actively encouraged forestry activity. The first step was legal and institutional restructuring aimed at increasing the area planted and encouraging industrial investment.

Simultaneously, the state became active in providing seedlings (produced in state nurseries) as well as in plantations themselves on both public and private land. From 1965 to 1973 (the year of the military coup against Salvador Allende's government) some 300,000 hectares of pine were planted. Similarly, the state participated directly in industrial activity, building a new pulp mill in Arauco (Celulosa Arauco) and starting another in Constitución (CELCO). Arauco was initially a joint venture between CORFO (80 per cent) and the US company Parsons & Whittemore (20 per cent). In 1972 CORFO (the state Corporation for the Promotion of Production) bought Parsons & Whittemore's shares and became the sole owner of the company. CELCO, meanwhile, was created by CORFO by the end of the 1960s, with shares also held by the Bishopric of Talca (ten per cent) and the French consortium Creusot-Loire Enterprises (18 per cent). The Bishopric withdrew from the project in a disagreement over the project's direction, and in 1974 CORFO bought the French consortium's shares and became the company's owner.

At this stage plantations constituted an important contribution to rural development, as particular attention was given to small and medium-sized producers, who reacted very positively to state initiatives. In contrast to those which were established later, pre-coup plantations did not take up the total area of any estate, part of which was set aside for crop and grazing land, and were owned by small and medium-sized producers. Labour conditions, too, were improving in the pre-coup years, as a result of marked development in union activity in the rural sector (Leyton 1986).

It is important to ask, however, why the state, technicians, businesses and rural producers alike concentrated exclusively on pine production, instead of attempting to manage native forests sustainably. After all, not only can indigenous forests achieve growth rates similar to those of radiata pine; their timber also fetches prices two to four times higher (Cavieres et al. 1983, Lara 1992, CODEFF 1992, Leyton 1986). Using native forests could also lead to important environmental and social benefits. The answer, according to CODEFF (1983), can be found in

> a style of development which, corresponding to external interests and markets, overlooks and undervalues existing forest resource potential, destroying with the aim of 'creating' the resources required by these markets. ... the reason for not using [existing forest] resources is not their lack of potential — that potential being quite high — but rather problems of commercialisation, lack of state incentives, and the nonexistence of a technological management model.

The dictatorship's model

The initial defects of the pine-monoculture model became more serious following the establishment of the dictatorship of Augusto Pinochet in 1973. Although the dictatorship advertised itself as following a 'neoliberal' model in matters of economic policy, this free-enterprise philosophy did not seem to apply to the forestry sector, where disproportionate state subsidies for export-oriented big business and direct investors became notorious (Lara and Veblen 1993, Leyton 1986).

The negative impacts of plantation forestry were aggravated and the positive ones minimized. As Leyton (1986) sums it up:

The two periods studied (1965–73 and 1974 to date) reveal divergent styles in relation to profit distribution and participation in the decision-making process. At present there is an extraordinary concentration of property, means of production, sales management and decision-making in the hands of a small number of businesses belonging to the three most powerful economic groups in the country. In return, there has been a considerable deterioration in conditions for the small proprietor and forestry workers who have remained marginalized from the benefits of growth in the pine economy. The current public policies have not favoured the wider strata of rural society, but they have to a large measure converted these into machines for the transfer of financial resources towards the big forestry companies of the radiata pine zone.

State assets of land, plantations and processing plants were rapidly sold at bargain prices to prominent actors in the private sector, granting them an enormous subsidy to help them compete on the international market. The state, which had taken the initiative and risks of long-term investment in forestry resources — for example, by planting 420,000 hectares of timber in the 1965–1973 period — handed over the fruits of its efforts to the private sector just at the moment when the returns on its work were about to start coming in (Gómez and Echenique 1988). In the words of Cruz and Rivera (1983), the 'current power of the large forestry businesses is based on public capital, as the handing over of the industrial infrastructure and plantations occurred at *artificially depressed prices*'. Large businesses have used the proceeds from this windfall to pay for tree planting and plantation maintenance. As Antonio Molina, president of the Rural Workers' Confederation notes, what large forest businesses earn today 'comes to them without having cost them anything' — and they, rather than rural workers, receive the benefits (*Chile Forestal* 1.1993). Between 1973 and 1979, moreover, the dictatorship returned around 4,000 of the landholdings expropriated during the agricultural reforms of the previous government to the earlier owners. This measure, which redistributed 28 per cent of all the expropriated land in the country, further encouraged large-scale forestry exploitation (Leyton 1986).

Similarly, it is not the actors who currently own the forest assets who had to put up investments in large pulp and paper factories, but rather the state, both directly through CORFO or through government support at the request of transnational corporations linked to national private projects (Leyton 1986). Some of the most significant agro-industrial conglomerates have benefited from this process:

- The present Celulosa Arauco y Constitución company, whose two pulp mills (Arauco and CELCO) were sold by the state in 1977 and 1979 to the Compañía de Petróleos de Chile (COPEC), the largest private firm in the country.

- Forestal Arauco, the seventh largest Chilean company, which owned 64,000 hectares of plantations in 1976 when it was taken over by COPEC.

- INFORSA, with a pulp mill, a paper mill and thousands of hectares of plantations, acquired from the state by Grupo Vial (one of the three major economic groups in Chile) in 1976 (Cruz and Rivera 1983).

Not sufficiently happy with their new property acquired at 'artificially depressed prices,' private industry pressured the state into providing a further set of incentives. These incentives ranged from direct subsidies for plantations (75 per cent of the cost), as well as for pruning, thinning, surveillance, and fencing, and even to guaranteeing the availability of a cheap workforce through the prohibition of union activity and repression of workers' and farmers' movements.

Within the confines of the current development model, which privileges the powerful sectors at the expense of the weakest, this policy has been highly successful. The Chilean forestry sector is today one of the main axes of the national economy, with pulp, paper and fibre shipments accounting for over five per cent of export income in 1991 (IIED 1995). The pulp industry is internationally considered to be a young giant (Swann 1993), well-positioned to sell to the growing Asian market, and the return to democracy has not caused major changes to government forestry policy (Lara 1992, Lara and Veblen 1993).

Support in implementing Chile's forestry policy has been offered by multilateral credit entities like the World Bank and the Inter-American Development Bank (IDB) and other multilateral agencies. Forest research and development, for example, have been promoted through the 'Investigación y Desarollo Forestal' funded by CONAF, the United Nations Development Programme (UNDP) and the Food and Agriculture Organisation (FAO). In 1991 a project to develop a Forest Action Plan for Chile was initiated as part of the worldwide Tropical Forest Action Programme, a World Bank and FAO-supported initiative which boosts forest investment and links the interests of transnational forest industries and consultants with those of Southern business elites and forest departments (Leyton 1986).

During the last few years, eucalyptus (particularly *Eucalyptus globulus*), has been incorporated into this model, mainly as a source of raw material for chemical pulp for export. In 1988, 8,000 hectares of eucalyptus were planted, and the yearly plantation rate increased to 17,000, 29,000, 34,000 and 41,000 in successive years. This increase was spurred by the good prices paid for pulp timber and wood chips, the fact that eucalyptus grows faster than radiata pine, and the fact that radiata pines have begun to suffer from disease, making them a risky investment. Eucalyptus plantations have replaced both native forest and arable and pasture land (Cerda et al. 1992, Cabaña 1993, Rada 1992).

Interest in eucalyptus on the part of Chilean and Japanese industries was awoken by the Santa Fé project (a venture involving Shell, Scott Paper and Citibank), which built the first eucalyptus pulp plant and simultaneously acquired the most extensive eucalyptus plantation in Chile (*P&PA* 5.1991). Instead of merely importing wood chips, Japanese firms became large-scale investors in Chilean eucalyptus plantations. It is estimated that in the coming years, subsidiaries of Itochu, Daio Paper, Mitsubishi Paper, Sumitomo Corporation, Nippon Paper and others will plant between 10,000 and 16,000 hectares annually, aiming at the yearly export of between 3.5–5.6 million tonnes by the beginning of the next decade. Similar planting rates are planned both by firms which have already built pulp mills (CACSA, CMPC, and Santa Fé) and by new groups proposing the installation of a new eucalyptus pulp plant (Andinos and Forestal Ace) (Rada 1992).

Socioeconomic and environmental consequences

Who benefits?

Most government support for Chile's forestry sector during the 1970s and 1980s went to only a few corporations. Four holding companies are estimated to own 40 per cent of all forest plantations and account for almost 70 per cent of forest exports. Another nine per cent of Chile's plantations are in the hands of seven holding companies controlled by foreign capital, which account for another 10 per cent of forest exports (Cabrera 1989, cited by Lara and Veblen 1993). Two of the large economic groups alone, Matte-Alessandri and Angelini, control nearly 50 per cent of existing Chilean pine plantations. Since 1975, moreover, 'forest corporations have been acquiring land from private owners, which, added to the large expanses acquired in open bidding from the state, has allowed them to concentrate enormous stretches of plantations' (Gómez and Echenique 1988).

In addition, the three most powerful Chilean economic groups (Vial, Cruzat-Larraín and Matte-Alessandri) have acquired not only plantations, the main pulp and paper mills, and, through subsidies, enormous forest holdings, but also sawmills, timber agglomerates, packaging plants, and other factories (Cruz and Rivera 1983). They are thus able to coordinate plantation operation with forest exploitation, industrial processing, transportation and sales (Leyton 1986).

Pulp production, concentrated in five large mills, is dominated by two large companies, Compañía Manufacturera de Papeles y Cartones (CMPC) and Celulosa Arauco y Constitución (CACSA), with significant participation by a third firm, Forestal e Industrial Santa Fé (Swann 1993). Arauco is owned by Compañía de Petróleos de Chile (COPEC), which is controlled by Chilean entrepreneur Anacleto Angelini. Angelini is a partner of Carter Holt Harvey of New Zealand, which is in turn jointly controlled by the US's International Paper Company and Brierly Investments of New Zealand — a firm which is also responsible, through a joint venture, for the illegal seizure and degradation, through eucalyptus plantings, of fertile public lands in Thailand (see chapter 12). Arauco operates the Arauco I and II and Constitución mills, with a combined capacity of around 850,000 tonnes per year, giving the company a three per cent share of the world pulp market (Swann 1993). In 1992 Arauco was the top national exporter, with an income of US$314 million (*PPI* 11.1993). CMPC, meanwhile, operates a pulp mill in Laja, with a capacity of 315,000 tonnes per year, and another plant of equal capacity in Mininco (Celulosa del Pacífico) in partnership with the US's Simpson Paper (*Papermaker* 8.1993). CMPC owns almost 415,000 hectares of land, most of which are planted with pine (*PPI* 8.1993).

Another firm, Forestal Santa Fé, is owned 60 per cent by Royal Dutch Shell, a British/Dutch firm, and 20 per cent each by the US's Scott Paper and Citicorp. Santa Fé runs a pulp mill with a capacity of 240,000 tonnes per year, of which Scott Paper (now encompassing Kimberly-Clark) is contracted to buy 40–80 per cent (Swann 1993). The Swiss firm Cellulose Attisholz meanwhile holds 25,000 hectares of *Pinus radiata* and another 100,000 hectares of land suitable for planting; all of its pulp production is exported to western Europe (*PPI* 11.1993).

The export orientation of Chile's forest industry and forest policy also benefits large paper companies in the North, who need increasing quantities of cheap raw material to maintain and increase current levels of consumption. Some 60 per cent of Chile's pulp production — mostly chemical pulp — is exported (Cerda et al. 1992). More than three-quarters of this goes to the North and to the East Asian 'tigers', with Europe being by far the biggest customer (FAO 1994).

As a result of the indiscriminate support on the part of the Chilean state for large national and international businesses, this pulp is one of the cheapest in the world (Shell/WWF 1993; *P&PA* 5.1991). According to a World Bank study, the production costs of pine logs (from planting to transportation to points of export) are only between 30 and 50 per cent of the normal costs in the US and the Nordic countries (cited in Messner 1993). Such low costs, as we will see later on, do not take into account the high social and environmental costs which underlie them. Chilean citizens, in other words, subsidize Northern consumers.

Chile is also an important exporter of other timber goods (Shell/WWF 1993). Of the over 400 forest products which Chile ships abroad, the mainstays continue to be pulp and unprocessed or semi-processed goods such as pulp logs, wood chips, and sawn timber. Principal markets are in Asia (especially Japan) and Europe, with South America and the United States also important (Cerda et al. 1992).

Who loses?

Chile's rural people are not only missing out on the benefits of the country's forestry 'development', but are by and large actually in a worse situation than before plantations were established. The current pattern of forestry development is even seen by many rural people as a threat to their survival (Otero 1990). One rural cooperative leader has suggested that the pine tree symbol on the current logo of Chilean rural cooperatives be removed, as this species is seen as an enemy by the local population (Leyton 1986).

For one thing, forestry activity, far from creating more employment, has directly caused the expulsion of small farmers and wage earners from rural areas. Censuses demonstrate that the areas with greatest plantation cover are those which have expelled the greatest number of rural workers, while at the other end of the scale, the agricultural areas of the central valley, which produce traditional crops, maintain their agricultural population (Leyton 1986). Even officially, it is recognised that 'poverty, even though it is an old problem, has been aggravated by large rural–rural and rural–urban migrations as a result of the expansion process of modern forestry' and that 'zones of high plantation concentration see the effects of enforced rural migration, with high levels of poverty and marginalization' (Chile 1991). After emigrating to unfamiliar rural areas, migrants must then often move on to cities, a process accompanied by increasing levels of alcoholism and prostitution (Lagos 1993a). Settlements called 'forest shantytowns' have meanwhile sprung up spontaneously on public ground, along the sides of roads, rivers or old railway lines (Cruz and Rivera 1983, Leyton 1986). The exodus stems from several causes:

- When forestry businesses acquire estates previously dedicated to agricultural production, most or all of the workers are swiftly dismissed. With trees planted on all available estate land, the rural population is left with few ways of making a living. One estate originally employing 260 people cut its staff to 14. On another estate, only one worker was kept on (as a guard) out of an original 120 (Leyton 1986, Equipo de Pastoral Campesina 1993, CODEFF 1994).

- The closure of many small sawmills and the concentration and modernization of the industry in a few large, highly mechanized mills has further reduced rural employment.

- Rural people seeking new forestry jobs must emigrate to the towns, since employers contract their workers there.

- Because forestry companies fear that local rural people may start forest fires, they also try to coerce them into emigrating. Firms kill domestic animals (Cruz and Rivera 1983); fence the countryside (Equipo de Pastoral Campesina 1993, Leyton 1986); cut off road access; and deliberately exclude local people from consideration for forestry work. As a result, many farmers and smallholders are forced to sell their land to the firms (Equipo Pastoral Campesina 1993). Very rarely, firms have been so eager to get rid of local residents that they offer higher-than-market prices for farmers' land (CODEFF 1994).

- The environmental impacts of large plantations have also directly led to emigration. In many cases, rural dwellers have found themselves deprived of water for themselves and their livestock. The replacement of native forests by pine and eucalyptus plantations has taken away other important components of rural residents' system of survival as well, including wood, fuel, fruit, fibres, dyes, honey, mushrooms, forage, game and medicinal plants (Otero 1990, CODEFF 1992). The application of herbicides and pesticides by forestry companies has meanwhile threatened the health of human beings and domestic animals alike (Cruz and Rivera 1983, Cavieres and Lara 1983). Large forest fires are another cause of displacement. In 1988, for example, over 18,000 hectares of plantation were burned in the Bío-Bío region , resulting in the destruction of 80 homes (Otero 1990).

Another reason for popular hostility to the new forestry model lies in the tough working and living conditions in the plantation and forestry sector faced by those who remain in the countryside.

Largely responsible for these conditions are post-coup changes in power relations between companies and workers. For example, the surplus of seasonal and unemployed labour brought about by the uprooting of rural people, coupled with the repression of the organized labour movement and the absence of almost any legal protection for labourers regarding salaries, working hours, working conditions, safety regulations, etc. (Federación 1988), has allowed big forestry companies to impose an employment system based on subcontracting. Under this system, forestry industries dismiss most of their permanent labourers, maintaining a reduced number of employees for administration and supervision (Gómez and Echenique 1988) and hiring subcontractors to carry out the bulk of the work. Only 2.3 per cent of the staff at Crecex, for example, is permanent, most of the firm's 2000 workers being dependent on seasonal contracts issued by subcontractors. Subcontractors, which are usually small enterprises, meanwhile compete strenuously for contracts by reducing their costs — mainly payroll — to a minimum (Leyton 1986).

The lack of worker unity promoted by the subcontracting system, combined with prevailing high rates of unemployment, big business' political power, and legal restrictions on unionization, has helped push forestry wages down to minimal levels. Unsurprisingly, the current forestry development model was implemented most swiftly during the fiercest part of the Chilean dictatorship following the 1973 coup. As the union movement was disbanded and a 'free' labour market developed, businesses were enabled to gain large profits by lowering salaries (Leyton 1986).

Seasonal labourers, in addition, lack job security and usually work without a contract, without protection against dismissal, and without rights to collective bargaining or old-age pensions. Informally-hired workers are dismissed as soon as their job is finished — usually a matter of weeks or months — and must return to unemployment until another job offer comes up (Federación 1988). José González Castillo, president of the Chilean National Confederation of Forestry Workers, estimates that 75 per cent of forestry workers are in precarious employment, with fixed-term contracts depending on the whims of the contractors (*Chile Forestal* 5.1993). In spite of the existence of an estimated 100,000 workers in the sector, the total number of days worked show that only half of these workers are employed on a permanent basis (Cruz and Rivera 1983).

The majority of subcontractors, moreover, do not provide their workers with safety equipment, even obliging workers to provide their own chainsaws. As Leyton (1986) notes, living conditions in the camps are on a minimum subsistence level. The housing has no plumbing, and often not even flooring. It is normally built from wood cut on site, with no sanding and lacking even the most basic comforts . . . the working day runs from six in the morning to eight at night on the plantations . . . a situation worsened by the *'pulpería'* system whereby the contractor sells all the basic goods the worker needs on site. These sales are made without money changing hands, the amounts being noted down and deducted on payday. These discounts can easily add up to 60 or 70 per cent of the worker's salary, as a result of the markup imposed by the contractor.

One unionist maintained that even this situation represented an improvement on the situation immediately following the coup, when 'the majority of us slept out in the woods all year round in the rain, dirt and cold.' It is hardly surprising that unionists describe subcontractors as 'twentieth-century slave dealers' (Federación 1988).

Despite benefiting hugely from this type of exploitation, large businesses cynically affect disappointment at the 'low productivity' of Chilean forestry labour. Responding to this 'concern', the University of Concepción, in a number of studies 'to determine if it is innate characteristics or social problems which explain the low productivity of the forestry worker', concluded that Chilean forestry workers have an aerobic physical capacity similar to that of Swedish forestry workers and better than that of most of the Chilean working population. The studies found, however, that because of poor diet and deficient sanitary conditions, Chilean workers only use 27 per cent of their capacity while their European counterparts use 50 per cent (Leyton 1986).

Environmental degradation

A study recently published by the Forestry Institute (INFOR) and the Corporation for Promotion of Production (CORFO) implicitly recognizes that Chile's current model of forestry development has so far tended to displace environmental damage onto other sectors:

> The increase in exportable timber on offer . . . and its consequent economic and social benefits allow both individual companies and the nation to offer compensation funds to *mitigate the environmental impacts of forestry and forest industries* (Cerda et al. 1992; emphasis added).

Three such 'impacts' will be discussed here: on biodiversity, on water, and on soils.

Biodiversity

One of the main effects of the new forestry on biodiversity, of course, derives from the conversion of large areas of native forests into industrial plantations of pine and eucalyptus through clearcutting or burning. Some 50,000 hectares of native forest disappeared in two of Chile's main forested regions (Regions VII and VIII) between 1978 and 1987, among which practically a third of the coastal forest in the VIII Region was felled in order to be replaced with pine plantations (Lara and Veblen 1993).

These conversions have threatened several plant and animal species, including three species of tree (*Nothofagus alessandri*, *Gomortega keule* and *Pitavia punctata*) and one shrub (*Berberidopsis coralina*) native to Regions VII and VIII, which are on the list of woody species in danger of extinction (Lara 1992). The plantations, whose density prevents the development of accompanying vegetation, also wipe out native plant communities typically containing between 20 and 158 species of vascular plants (Lara and Veblen 1993, Schlatter and Murúa 1992).

Animal species have also been devastated. In the unmanaged pine plantations of the central zone, the ground is covered by:

> ... fallen needles of an orangey coffee colour. From time to time fungi can be spotted — especially after a rainstorm — along with various other small plants. The fauna is very scarce; the forests lack birdsong and have no amphibian life. There is a total absence of reptiles and mammals. Only in the fringe zone, the fire breaks and open glades are there greater signs of life (Schlatter and Murúa 1992).

According to the National Forestry Corporation, the monoculture tree plantations constitute a threat to the survival of various endangered species like the pudu (Chilean goat), the Chilote fox, the little long-nosed weasel, the huemul (Andes deer), the little forest monkey and the Darwin frog, among others (CODEFF 1992).

Pulpwood plantations, of course, are hospitable to certain species. But the homogeneous nature of the plantations allows such species to run rampant in them, sometimes resulting in the annihilation of whole stands of trees. For example, two species of field mouse, accustomed to feeding on roots, recently adapted themselves to eating pine as a result of the changes in their environment. In some zones this resulted in the destruction of up to 30 per cent of the trees. In the last few years, similarly, *Rhyacionia buoliana* (European pine shoot moth) has also appeared in Chile, eating the internal sections of terminal pine shoots and forcing the tree to put out new ones, causing a loss in growth and directionality, weakening the tree and leaving it open to fungus infections which eventually kill it. The fungi *Diplodia pinea* (which mainly kills the tip of the tree) and *Dithistroma pinea* (which causes the needles to fall) are other serious afflictions of pine plantations, while *Bacunculus phyllopus* (an insect) eats pine needles and hinders photosynthesis (Otero 1990). Epidemics of such organisms pose serious problems for local people who have become economically dependent on forest production (Schlatter and Murúa 1992).

Other species can also become problems. The planting of pine in monoculture, by altering local vegetation structure, has 'prevented nest building by birds and stopped foxes from being able to move around inside the forest. Rodents and rabbits, competitive species, have increased, endangering young pine plantations and causing economic losses to forestry companies' (Schlatter and Murúa 1992).

The companies have then applied chemicals which poison not only the rabbits, but also other mammals and birds. Carnivores needed to regulate the herbivore population go into decline as a result, creating yet more opportunities for the rabbit population to increase and promoting a vicious circle (Cavieres and Lara 1983). Herbicides (including one containing a component of Agent Orange) have also been used on young plantations for weed control, and undoubtedly cause further declines in indigenous biodiversity (Lara 1992).

The use of fire as a management method has also reduced biodiversity. As Cavieres and Lara (1983) note, fire

kills practically all the existing fauna in the area in all stages of development: eggs, young, adults etc. . . . Not only are whole populations of various species killed, but also, through actions of this type, their habitats and micro-habitat are destroyed (burrows, nests, etc.), a situation which persists for a long time or even indefinitely in some cases.

In the province of Bío-Bío, the copihue (the national flower of Chile), has seen its 'presence in the study area seriously diminished as a result of the fires, with only a few scarce weedy examples now existing'.

The process of substituting plantations for native vegetation also impoverishes the landscape as a whole. A diversity of life, whose outcome is a landscape of unique characteristics, is transformed into the monotony of uniform lines of one species of pine. A recent study has shown that the pine plantations are less attractive to tourists than areas of native forests (CODEFF 1992).

Soil

With each new study, evidence accumulates that industrial plantations are an important factor in soil degradation. This degradation can be attributed to several factors.

First, unlike native forests, pine plantations tend to extract more nutrients than they give back to the soil, due to the absence of a rapid humus-production process. They thus gradually reduce soil fertility (Gayoso, cited in CODEFF 1992). The rapid growth of the Chilean plantations, in other words, is owed, 'among other reasons, to the existence of a large quantity of nutrients generated by the native forests'. It is this which has made it possible for forestry firms to plant radiata pine compactly and achieve better growth rates than in this pine's original environment (Cruz and Rivera 1983).

Poor humus production is in turn a result of several factors, especially the soil acidification associated with pine plantations, which impedes the development of microorganisms which help break down organic materials. While this acidification simultaneously encourages the growth of a different type of microfauna, especially various forms of fungi that help pines absorb nutrients from the soil, these fungi are not capable of producing humus (Cruz and Rivera 1983). The faster the pines grow, the more impoverished the soil becomes.

As one academic apologist for plantations admits, moreover, two or three rotations of pines lead to a 'pronounced loss of some nutrients, like boron, making fertilization necessary'. Pine plantations thus have to be treated, even in the eyes of their defenders, as 'an intensive crop and have to be dealt with technically just as agriculture has its annual crop management' (Cruz and Rivera 1983). Pine plantations, in other words, are the forestry equivalent of Green Revolution agricultural crops, the negative impacts of which have been extensively described by Vandana Shiva (1991a).

Erosion is another source of soil degradation on plantations. The felling of native forests and the burning of residual wood leave the soil unprotected for the first two or three years of a plantation, resulting in intense erosion during the heavy winter rains (Lara 1992, Cavieres and Lara 1983). The same thing occurs after the final harvest, when all the plantation trees are felled and the remains burnt off. As Otero (1990) observes,

> Studies by CONAF in the VII Region show that the practice of burning the waste, as is currently done in over 10,000 hectares of the zone, causes a loss of between 35 and 566 tons of topsoil per hectare per year, in circumstances where the maximum tolerable loss for forest soils should not be above four tons per hectare per year.

This process is even more serious in mountainous areas, Otero continues, where the burning of residues leads to the loss of between 500 and 2,000 tons per hectare per year. When a forest is exploited, moreover, access roads and the dragging of logs open deep ruts down which winter rains flush topsoil and undecomposed needles (Cruz and Rivera 1983).

Plantation supporters, citing the insignificant soil loss on certain *unmanaged* plantations, have sometimes concluded that plantations are the most efficient method of recovering soils (Endlicher 1988, cited by Otero 1990). Such data are irrelevant to the majority of Chilean pine plantations, which are commercially managed and are thus oriented toward timber extraction at the fastest possible rate.

Water

Chile's current model of forest development has, in the words of Leyton (1986), contributed to 'levels of extreme danger in terms of flooding or scarcity of water on a local level'. Industrial plantations (and associated industrial processes) affect water supplies in two ways: through changes in the hydrologic cycle and through contamination.

A large body of evidence demonstrates that pine plantations cause important changes in the cycling of water in catchment areas. According to studies carried out in 1991 by Anton Huber (cited in CODEFF 1992), an adult radiata pine tree evapotranspires — liberates into the atmosphere — 60 per cent more water than adult native trees.

In addition, the soil under pine plantations has little rainfall absorption capacity, with a thick layer of undecomposed needles often preventing water from even reaching it. As one forest ranger in an area of extensive pine plantations has observed, the humus layer of a native forest is always damp and a great number of small sources of running water can be found there. The floor of a pine plantation, on the other hand, is dry for most of the year. Even after heavy rains there is little noticeable dampness, and that is seen only on the surface of the carpet of needles; ten centimetres down the soil is dry (Cruz and Rivera 1983). As the Huber studies concluded,

> [P]ine plantations, in continuous extensive stretches, cause a great desiccation of water sources. The smallest streams shrink or disappear, and as a result the flow of the larger water courses and the water supply for rural populations do likewise.

In some areas, streams dried up when plantations were established, reappearing only when the trees were felled. Wells, too, have dried up during the summer months, depriving local residents of water both for themselves and for their livestock. Towns such as Angol began facing serious water supply problems eight years after plantations were established in the surrounding countryside. In some cases houses have been abandoned when water supplies dwindled. The inability of soils under pine plantations to hold water, combined with a lack of water-retaining undergrowth, can also lead to flooding in the valleys below, when, after storms, rainwater rapidly flows down hillsides (Cruz and Rivera 1983).

Increasing use of fertilizers, herbicides, and pesticides on the part of the forestry sector, moreover, is unquestionably leading to the contamination of watercourses. Erosion has also caused high levels of cloudiness in water supplies (Otero 1990), sometimes making them unfit for drinking (Cavieres and Lara 1983). The downstream industries fed by tree plantations, of course, also pollute water supplies. As a result of a pulp plant, for example, the tourist city of Constitución has seen its beaches contaminated and its coastal fisheries damaged (Cruz and Rivera 1983).

Conclusions

Present-day Chilean plantation forestry constitutes an example of socially regressive and environmentally unsustainable development. Arising under a military dictatorship which offered its support to large economic interests and transnational capital, this style of development has engendered poverty, despoliation and exploitation. Not only have the country's plantations failed to ease pressure on native forests, improve soil and water conservation, or promote employment and social development; they have also brought about an increasing artificialization of the ecosystem, particularly through the use of agrochemicals which create even greater problems than those they claim to solve.

What is extolled as an ingredient of a 'boom' in exports and industrial production, moreover, has resulted in the decline or even disappearance of much of Chile's rural society. As Badilla (Equipo 1993) puts it, 'History is repeating itself. In the past it was the indigenous peoples who suffered the rape of their lands; and now we are seeing the same with the rural population'.

While a factory model of monospecific industrial plantations is being promoted, native Chilean forest resources, which are of benefit to the majority of both present and future Chileans, continue to be ignored or pillaged, since managing them in a sustainable way is not in the economic interests of a small, powerful minority based both in Chile and abroad. To Chile's rural people, the advance of pine and eucalyptus plantations constitutes the 'advance of a green army' (Equipo 1993).

Chapter 9

Uruguay: 'Forests' on the Grasslands

The Uruguayan landscape is dominated by prairie ecosystems. Over 90 per cent of the territory is grassland, a fact which led the Spanish conquerors to introduce cattle, which reproduced quickly and successfully on the country's fertile plains.

At the time the Spanish arrived, what native forests existed were mainly found bordering the many rivers. This is still true today, although the forests have dwindled and the variety of species has decreased, with some species even disappearing in some areas. At present forests cover three per cent of the national territory (some 600,000 hectares), down from an estimated original figure of about six per cent.

From tree planting to industrial plantations

Today Uruguay is embarking on intensive forestry, mainly the plantation of large masses of eucalyptus and, to a lesser extent, pine. Because these swathes of exotic trees are being planted on top of prairie ecosystems, where, by and large, forests never previously existed, this process clearly does not constitute reforestation. On the contrary, as an ecosystem change, it amounts to an environmental disaster similar to the internationally-condemned replacement of tropical forests by grasslands, except that here it is the prairie which is being destroyed.

Tree plantings began with the arrival of the first Spanish colonists, who brought with them seeds of pine, olive, poplar, oak and many other exotic species, which were grown mainly for ornamental reasons. Later, tree seeds from even more distant regions were introduced, among them various species of eucalyptus in 1853 and *Pinus pinaster* in 1890, both of which became staples of 20th-century plantations.

The initial plantation model might be described as one of 'spontaneous agroforestry systems'. In general it involved the plantation of small patches of eucalyptus on cattle ranches for the purpose of providing shelter and shade for the livestock which constituted the economic basis of the nation. Because the main aim was not timber production, no intensive management techniques were used. In general the plantations were made up of a mixture of various species of eucalyptus (*Eucalyptus globulus, E. tereticornis, E. camaldulensis,*

E. saligna, E. robusta, E. diversicolor, E. cinerea, and so forth). No prior preparation of the soil was required (the trees were simply planted in holes), nor were fertilizers used, although some chemicals to combat leaf cutting ants were applied during the first two years of growth.

These plantations had no significant negative impacts. Their small extent minimized their effects on local hydrological systems and on flora and fauna. Furthermore, the main tree species used did not turn out to be invasive of other ecosystems, and the use of a variety of species protected the plantation from pests and diseases. Only *Eucalyptus globulus* (the species most often planted in solid masses) suffered disease, being attacked extensively by the weevil *Gonipterus gibberus*. Soil was not lost, nor nutrients exported; when the plantation reached maturity, a balanced nutrient cycle was achieved. Low chemical use, in addition, prevented contamination of water.

This plantation model generated confidence in eucalyptus plantations generally; indeed, today many Uruguayans believe eucalyptus to be an indigenous tree. At the same time, it provided experience in seedling production, plantation and plantation management techniques, and the best species were identified for various soils and regions.

As this plantation model developed, another was also being explored: pine plantations to stabilize the dunes lining the beaches in Uruguay's south. These plantations, copied from those established in southwest France (*les Landes*), were in line with other contemporary efforts to dominate nature. Even though the experiment ended in the near-elimination of the original ecosystems, one important by-product was the development of an economically very significant tourist industry which today draws visitors from foreign countries as well as from Uruguay itself. As with eucalyptus, the pine became 'naturalized' in the Uruguayan consciousness and its planting therefore tends not to raise resistance among the public in general.

These two models began to change as a result of an economic crisis which began in the late 1950s and forced the country to turn from foreign to domestic wood sources. It was then that the potential of wood supplies from self-regenerating pine plantations was discovered. Timber buyers also began to obtain supplies from the eucalyptus plantations on livestock ranches. When international oil prices rose in the 1970s, too, the manufacturing sector found in eucalyptus a far more economical and, initially, abundant source of fuel (gasified wood). Plantings increased, helped by a 1968 forestry law which, through tax exemptions, promoted species suitable for the timber industry. However, it was not until the military dictatorship of 1973–1985 expanded these exemptions substantially that large areas were finally planted, between 1975 and 1979, with eucalyptus — mainly for fuel — and pines. In 1979 this subsidy was eliminated and new plantation development came to a virtual halt.

Toward the end of the 1980s the country moved on to the export model, according to which 'forestry management does not differ much conceptually from the management of any business or factory' (Pou 1992). The aim was to produce for foreign markets at a competitive price using the technological package of the Green Revolution: intensive soil preparation; large scale monocultures of fast-growing species; fertilizers, herbicides and pesticides; and mechanized harvesting. The exportable product was initially to be eucalyptus roundwood for pulp, later on changing to pulp once a 'critical mass' of plantations was reached and pulp plants became feasible.

Domestic conditions encouraging the industrial model

The fact that Uruguay, originally lacking an industrial wood resource, has become an exporter of wood to countries as distant and traditionally as thickly-forested as the Nordic nations is very nearly beyond belief. It can be explained only by reference to an extraordinary set of conditions, partly internal to the country and partly external.

One domestic factor leading to the extensive growth of exotic tree plantations in Uruguay involves the country's cultural dependency. Uruguay, a nation of immigrants, has always looked at itself as a mirror of Europe and this vision is also reflected in its forestry practices. Instead of stressing that the country is covered by magnificent pastures, many Uruguayans have emphasised its lack of European-style forests as a negative element. For example, one of the first books on forestry edited in the country (Lopez and Cussac 1943), points to an 'insufficiency of forests in Uruguay', saying that Uruguay is 'by far the poorest nation in forests in Latin America', a state of affairs which 'brings with it serious economic, climatic and soil problems'. This point of view, shared by most technocrats and the public in general, helps explain the support tree plantation has historically received in the country.

Another relevant internal condition has been the stagnation of Uruguayan agriculture, due to a large extent to the constant fall in the international prices of its traditional products — meat, leather, wool and cereals. Wood is currently seen as a more dynamic agricultural activity. On a more general level, the nation is being increasingly pushed to increase exports in order to pay off a large foreign debt. The global pulpwood market is viewed as insatiable and therefore well able to generate the necessary foreign currency.

At the same time, Uruguay attracts transnational corporate forestry investment for several reasons. For one thing, land is cheap, fertile and relatively flat, favouring mechanized forestry operations. Land titles are clearly defined and socially accepted. In addition, climatic and soil conditions favour the rapid growth of the species needed for the paper industry.

Opposition in the thinly-populated livestock-raising areas where the government promotes plantations, moreover, is minimal. Indeed, the lack of other job opportunities in these areas encourages expectations that plantations will result in net employment gains. Rural underemployment helps depress wage rates to less than US$2 per hour (which Forestry Department leaflets aiming to attract foreign investors claim falsely to be 'one of the lowest levels in the world' [Uruguay MGAP 1994a, 1994b].)

In addition, the country has built up considerable technical expertise in plantation management, and basic infrastructure for extraction and export of forestry products already exists — though enormous additional investments will be needed to deal with the huge production levels that are foreseen. Officials project that 2.8 million tonnes of roundwood will be exported in the year 2001 and six million tonnes in 2004 (Buxedas 1995). This implies one loaded 30-tonne lorry arriving at a port every 15 minutes year round, or seven shiploads per month (Pérez Arrarte 1993).

A new forestry law (1987) to promote plantations with direct subsidies, tax exemptions and soft loans is also in place. The new law differs from the 1968 law mainly in offering plantation financing through a World Bank loan. Finally, Uruguay is attractive to foreign investors in that there is little likelihood they will be accused of destroying indigenous forests: not only are such forests a relatively small feature of the landscape, but they are also explicitly protected by national law.

External conditions

To take advantage of these long-standing domestic conditions, investors also required outside stimuli.

One long-term source of such stimuli has been the United Nations Food and Agriculture Organization (FAO). In 1951, a joint FAO/World Bank mission made recommendations which, taken together with those of a later mission (Rogers 1955), formed the basis of the forestry laws passed in 1968 and 1987. The vision of Uruguayan forestry development which FAO articulated at the time has exerted a great deal of influence over the country's foresters, who to this day repeat unsubstantiated FAO statements about the supposedly protective properties of plantation 'forests' — for example, that 'they will help to resolve ... the extremely important water supply problem' (Hutton and Winkelmann 1953).

Other international organizations such as the United Nations Development Programme (UNDP) and the Organization of American States (OAS) have also backed the industrial plantation model. The OAS, in particular, has contributed studies of markets, production, transport, industrialization, and so forth. OAS has been the only international organization to raise questions about the environmental sustainability of plantation forestry, stressing the need for research and for producers and technicians to maintain 'an open posture on the issue' (OAS 1994).

The principal external catalysing agents for industrial plantation development in Uruguay, however, appear to be the Japan International Cooperation Agency (JICA) and the World Bank. In 1986 a JICA team took up residence in the country for three months, working alongside Uruguayan government officials. On the basis of the team's studies, JICA published, in March 1987, a 'Study Report for a Master Plan for the Establishment of Tree Plantations and for the Use of Wood Planted in the Eastern Republic of Uruguay'. In December of the same year, the government passed the new forestry law, followed the next year by a national forestry plan explicitly based on the master plan put forward by JICA. JICA continued to participate in Uruguayan forestry thereafter, using staff permanently based in the country and new missions sent to study specific issues, though Japanese capital has not yet been directly invested in plantations or wood-related industries.

While JICA gathered and analysed much useful information, it restricted itself mainly to examining the economic viability of industrial pulpwood plantations and a pulp mill. The convergence in interests between the Japanese researchers and their Uruguayan counterparts — who were exclusively foresters — made it almost inevitable that the study would neglect non-plantation alternatives in rural development. The fact that the studies were seen to be Japanese gave the industrial plantation model additional credibility with Uruguayan investors and the Uruguayan public, as well as with foreign investors and the World Bank.

The JICA studies — which, incidentally, could have been done entirely by local experts — failed completely to investigate plantations' socioeconomic and environmental impacts. This irresponsible attitude is exemplified by the statement one JICA expert, Takahito Mikami, offered at a 1991 conference with the preposterous title 'Afforestation: Foundations for Building a Nation'. Contradicting available evidence, Mikami glibly assured the audience that plantations would offer 'advantages like regulating the climate, protecting against flooding, conserving water resources, fertilizing the soil and improving the environment' (Mikami 1991).

In the final report for the 1987 Master Plan, JICA's arguments were equally lightweight. Evaluating the functions of afforestation, the agency asserted without substantiation or explanation that 'the public functions of the forests are to guarantee water resources, prevent soil erosion and sand displacement, and provide protection to the natural environment'. In relation to water, JICA states, again with no evidence, that 'it is supposed that afforestation improves the water reserve capacity by 1,000m³/h', going on to recommend the afforestation of one of the nation's main river basins where three dams already exist for hydroelectric generation (JICA 1987). This recommendation seems irresponsible, to say the least, in the absence of previous studies, given the huge water consumption that fast-growing industrial tree plantations entail and the impacts they might therefore have on the water needed to feed the hydroelectric turbines.

The resources needed to back JICA's plantation vision were supplied largely by the World Bank. Indeed, it was World Bank finance that made the 1987 national forestry law into the effective tool for plantation development that the 1968 law had failed to become. In 1989, the Bank approved a US$65 million loan for the Second Agriculture Development Project. Some $27 million of this was aimed at helping to create 100,000 hectares of eucalyptus plantations for a short-fibre export kraft pulp industry and 60,000 hectares of pine plantations for both a sawmill and a chemi-thermo-mechanical export pulp industry. This 'critical mass' of plantations, to be planted over eight years, would also make possible both continued pulpwood exports and continued production of firewood (to consist increasingly of forestry residues) for the domestic market. Plantation enterprises were to be promoted in three ways: by developing wood products exports, by expediting the handing out of the financial incentives provided under the 1987 forestry law, and by providing credit for planting, logging and small industrial development (World Bank 1989).

The World Bank money had immediate effects. Both national and international entrepreneurs began to invest in the sector, with the latter in particular boosting its perceived viability. According to the Forestry Department, annual private sector afforestation in the 1989–92 period was eight times greater than average annual afforestation between 1979–88 (Uruguay MGAP 1994) and annual plantation has continued to increase at a similar rate since (see Table 9.1).

The most important foreign enterprise has been Forestal Oriental SA, owned 60 per cent by Royal Dutch Shell and 40 per cent by Finland's Kymmene (now merging with United Paper Mills to create Europe's biggest paper firm). Forestal Oriental is establishing 30,000 hectares of eucalyptus plantations to produce export timber for UPM-Kymmene mills in Finland. Drawing on the international experience of its partners (in particular Shell's failed plantation attempt in Thailand, where opposition from local people was strong), Forestal Oriental is currently the only company in Uruguay with a 'green marketing' policy. As a result, it carried out an environmental impact study for its plantation even before this was made a legal obligation under national law — though only a short summary has so far been made available to the public (Forestal Oriental 1993).

TABLE 9.1
Area of plantation established
per year, Uruguay (hectares)

Year	hectares
1975–88	2,045 (average)
1989	6,239
1990	10,791
1991	15,451
1992	23,631
1993	34,808
1994	33,768

Source: Buxedas 1995.

Spanish and Chilean businesses have also invested in Uruguayan plantations, encouraged by the President of the Republic himself, who recently wrote in a promotional pamphlet: 'With the experience of having invested in my own ranch, I recommend that you look into the opportunities and follow my example' (Uruguay MGAP 1995).

Voluntary blindness

Although current plans call for 200,000 hectares of tree plantations to be established in the five years from 1991, none of the multilateral or bilateral organizations supporting the plantations has studied their possible environmental impacts.

For example, Jaime Latorre, the Executive Director of the IBRD Forest Development Project in Uruguay, has denied that tree monocultures have any negative environmental effects — even though his project has carried out no research into this issue. Latorre claimed in 1991 that with the new plantations 'the country would be showing the world that establishing forest where there was none before is to become a member of the short list of countries who effectively contribute to environmental improvement' (Latorre 1991). The World Bank itself, in a report submitted to its Executive Directors, assumes without evidence that industrial plantations, far from bringing about environmental problems, will help counteract soil erosion and degradation (World Bank 1989). Although the OAS, by contrast, has suggested investigating 'how expanding forest plantations behave and how ecosystems respond' (OAS 1994), no research has resulted.

It has thus fallen to non-governmental organizations (NGOs) and a few independent researchers to study and disseminate information about the actual and potential social and environmental impacts of plantations. NGOs have presented enough evidence to justify state support for continuing research on the matter, with possible impacts on water and soils documented by Panario et al. (1991); on fauna by Gudynas (1989, 1990) and Gudynas and Rudolf (1987); and on society and the economy by Carrere (1989, 1992), Carrere et al. (1995), Pérez Arrarte et al. (1993), Damiani (1990), and Stolovich (1995).

Information countering official forestry propaganda has been disseminated to the public through a variety of activities. NGOs have used books and articles to raise consciousness about plantations, and workshops, seminars, conferences and public debates have been held with the participation of environmental NGOs, trade unions, agronomists, forestry company representatives, youth organizations, universities, and local communities in forestry areas. The different vision put forward by NGOs is allowing people to take a more informed view of plantations. As a consequence, opposition is gaining strength.

Interestingly, transnational corporations are more receptive than local foresters to the idea that industrial plantations may have negative effects on the environment. Environmental impact assessments carried out by Forestal Oriental and by the consulting firm Informes y Proyectos S.A. (INYPSA) for Transpapel (a paper mill project reportedly backed by Swiss capital) point to possible impacts on water, soils and biodiversity. Such warnings have not been taken seriously or have been greeted with silence by government officials and foresters in general. In June 1995, Roberto Cal, the Director of the Renewable Resources Department, said that campaigns by Uruguay's Environmental NGO Network pointing out environmental problems with plantations were 'based on erroneous information' and that in Spain 'it was shown that negative impacts did not exist' — an astonishing statement, given the work of Bermejo (1994, 1995), Groome (1988), González Bernáldez et al. (1989), Castroviejo Bolibar (1985), Márquez Fernández (1985), Ruiz Pérez (1990), PSOE (1979), Coordinadora Extremena de Proteccion Ambiental (1992), and others. In Uruguay, Cal maintained, 'there are no indications that . . . plantations provoke any negative impact or degradation' (*La República* 7.6.1995).

Yet ample evidence already exists of the intrinsic weaknesses of monoculture industrial plantations in Uruguay. Although initially successful extensive plantations of *Pinus radiata* were established in the 1960s (just as FAO expert Lewis Rogers had recommended), within a few years the plantations were attacked by the European pine shoot moth *Rhyacionia buoliana*, which, along with an associated fungus, *Diplodia pinea*, spread rapidly to nearly all the plantations. As a result, plantings of this species had to cease (Morey and Porcile 1992).

Fortunately, the country had not yet embarked on an ambitious forestry plan, so economic losses were comparatively small. Today, however, with 200,000 hectares already in place, in addition to the 200,000 to come, losses could be enormous, particularly since the plantations consist mainly of two species of pine (*Pinus elliottii and P. taeda*) and two of eucalyptus (*Eucalyptus globulus and E. grandis*).

Indeed, most current species already have pest problems, including insects such as leaf cutter ants of the *Atta* and *Acromyrmex* genera, *Platypus sulcatus* (a eucalyptus wood borer), *Pissodes castaneus* (a pine weevil), *Sirex noctilio* (the sirex wasp, which infests pines), *Gonipterus gibberus and G. scutellatus* (eucalyptus weevils), *Phoracantha semipunctata* (a longhorn beetle) and fungi which attack pines (*Diplodia pinea, Dothistroma pini*) and eucalyptus (*Phytophtora cinnamomi, Cladosporium eucaliptii, Septonema eucalipticola, Pestalozzia molleriana and Alternaria* spp.). As Morey and Porcile (1992) state, 'increased forest resources entail a potential increase in pests and diseases along with a greater risk that new damaging organisms may find conditions more favourable for their expansion in the country'. And if chemical pesticides have to be applied every time that a pest cannot be controlled by silvicultural means, serious environmental consequences are likely to follow.

Small farmers, moreover, have seen their yields damaged by plantations. In the period from the mid-1970s to the mid-1980s when plantations were being designed to meet fuel needs, small producers were extremely active in trying to get a ban on planting eucalyptus in predominantly agricultural areas, although their reasoning was not fully assimilated at government level. Among other effects, the small producers mentioned crop pests, competition for water and nutrients, shade-out effects reducing agricultural yields, and the danger of fire. The latter, it was felt, even led to personal insecurity due to the proximity of plantations to farmers' homes (Damiani 1990).

Land tenure is also being affected by plantations. Before the plantation boom, large cattle ranches had been decreasing in size, mainly as a result of inheritance laws. Now, however, plantations are bringing about an agrarian reform in reverse. Investment in the forestry sector is allowing land to be concentrated in fewer, and, in many cases, foreign, hands. The typical Uruguayan *estancia* or cattle ranch covers 2–3,000 hectares, ranches of more than 10,000 hectares being extremely rare. Yet the ventures dominated by the transnationals Shell/Kymenne and ENCE, a Spanish pulp and paper company, already own more than 30,000 hectares each, and are planning to double their landholdings. The new landowners, in addition, are typically investors and not traditional farmers. This could lead to important changes in Uruguayan society, with an increased dependence on new transnational actors who are likely to come to dominate large areas of the country.

Rural workers, meanwhile have little to gain except a few jobs, and these are characterized by low pay, instability, seasonality and poor working conditions — particularly where contractors are involved (Carrere 1989, Stolovich 1995). Yet even these jobs are in some places preferable to being unemployed or living off the low wages on offer in the cattle-raising sector. Although conditions vary from place to place, moreover, the level of rural unionization is extremely low due to unions' limited experience, low worker concentrations, seasonality of work (half of all forestry workers work on a seasonal basis), workers' fears of losing their only possible employment possibilities, and in some cases repressive management (Latorre 1995). Unions exist only in the few areas with liberal management regimes (for example, those of companies managing forest investments for bank workers' retirement schemes) and large concentrations of full-time workers. The solution, as sociologist Raúl Latorre (1995) suggests, is not plantations but rather 'a state policy which, in addition to incentives for investors, offers measures to ameliorate workers' quality of life'.

A bad investment for the nation

One of the main claims made by the government for tree crops is that they will generate substantial foreign exchange. However, analysis of the probable export figures casts doubt on this assertion. When all costs are included, the balance could even turn out to be negative.

Pulpwood exports are likely to net the country only US$53 million in income by 2000 and US$82 million by 2010, or around three and four per cent of total annual national exports (Stolovich 1995).

TABLE 9.2
Uruguay pulpwood exports: gains *vs.* losses for the country as a whole in two indicative years ($US million)

Category	2000	2010
Year's pulpwood exports	+53.1	+81.8
Year's subsidies and tax exemptions	–10.0	–12.0
Year's infrastructure expenditures specifically for pulpwood transport	–45.0	–45.0
Year's import costs	–6.0	–7.9
Year's opportunity costs	–14.4	–21.4
Balance	*–22.3*	*–4.5*

Source: *Stolovich 1995.*

But even to achieve these levels of exports, the country will have to take resources from other sectors and pour them into forestry. Tax exemptions, highways, bridges, railways, port facilities, state investments in plantation and machinery imports — all must be paid for by the public. This is not even to take into account the income lost when land is taken out of other forms of production and planted to fast-growing trees. Economist Luis Stolovich (1995) has sketched out a balance sheet showing that financial gains from projected pulpwood exports in the year 2000 would be far outweighed by losses and that the balance would still be negative in 2010 (see Table 9.2). As one NGO analyst puts it, 'the export of roundwood for pulp . . . means going back to the 19th century':

> Uruguay would not only become an exporter country of commodities with little added value, but would also be making a great social effort (of social transfers, investments, etc.) to promote an activity which, even in the best case, would make few social contributions and would not be a source of surplus to fund a global economic revitalization of the nation (CIEDUR 1991).

According to Stolovich, the balance becomes positive only when pulp exports are also included. In that case, a projected surplus of US$20 million is generated for 2000 and US$60 million for 2010.

Even in current dollars, however, this is not very substantial in comparison with Uruguay's 1994 external debt of $5.253 billion (World Bank 1995). Stolovich's figures, moreover, do not include environmental costs such as the impacts of plantations on water, soils and biodiversity as well as industrial pollution. They also assume that the country will be able to export all its pulpwood and pulp, a dubious premise given that in 1993, for example, 'exports came to a near-standstill because the international price for pulpwood did not allow export dealers to buy wood internally' (Pérez Arrarte 1994). As Stolovich notes, specialization in commodities such as pulp, which may well fall in price in the long term, is likely to put the export sector 'in a state of permanent uncertainty at the mercy of concentrated demand' (Stolovich 1995). Such uncertainties and costs only add weight to the basic question of why the Uruguayan people have not been given the opportunity to decide whether so many of the the country's scarce economic resources should be spent on pulpwood plantations.

Conclusions

The successive Uruguayan governments which have promoted an industrial export model of forestry development claim to be following a neoliberal economic policy and thus to be denying all forms of subsidy to productive activities, leaving them to their own devices in an increasingly globalized and competitive international market. In truth, however, they are pouring resources into the sector. While withholding funding from education, social security, health, housing, and so on, the government is subsidizing giants like Shell, 'which, worldwide, move figures far greater than the whole national economy' (Stolovich 1995). Without such subsidies, this form of forestry development would not be profitable for private investors. Even with them, it cannot substantially improve the country's export earnings.

Industrial afforestation for export, moreover, not only does not improve the environment, but also threatens to degrade it. Plantations supplanting grasslands have a deep impact on local and regional ecosystems and their soils, water, plants and animals. At the social level, the employment opportunities created do little to improve the quality of life of rural people, at the same time that land and other forms of power are concentrated in fewer and fewer hands. In short, while the new industrial plantations will benefit Northern-based paper industries, who will be enabled to exploit a fresh source of cheap and homogeneous wood fibre, heavy environmental and social costs will be paid by present and future generations of Uruguayans.

Chapter 10

South Africa:
A Fibre Exporter with Few Forests

South Africa's native forests, which probably never occupied more than one per cent of the surface area of the country (Cooper 1990), shrank significantly following the arrival of European colonizers, mainly due to overexploitation in the 19th century. Today between 200,000 (Bethlehem 1994) and 300,000 (Dudley 1992) hectares remain, and these continue to be degraded — despite the fact that the majority are within protected areas — as a result of poor land use strategies (Dudley 1992).

Though South Africa lacks extensive natural forests, the country has become a significant exporter of forest products from large plantations of pine and eucalyptus established as the result of a long process of state intervention. The first plantations date from 1890 (Bethlehem 1994), though eucalyptus had been introduced as early as 1807 (FAO 1981). Between 1890 and 1900, most plantations were small and were aimed at satisfying the needs of local communities.

Large plantations began to appear at the beginning of the 20th century, established mainly by the state and aimed at the production of timber for industry. Wood shortages resulting from supply difficulties during the First World War encouraged the government to embark upon more intense tree plantation programmes, initially to achieve self-sufficiency in timber. The post-war depression, along with unemployment in the poorer sections of white society and among soldiers returning from the battlefields, gave the government additional incentives to undertake large-scale afforestation (Bethlehem 1994). The main species used were exotic pines (particularly *Pinus patula* and *P. radiata*), eucalyptus (mainly *E. globulus*) and Australian acacias. By 1923 the country boasted 140,000 hectares of acacia, 65,000 of eucalyptus and 35,000 of pine (van der Zel 1990).

From 1920 to 1960, the state became the main national forester, with private industry occupying a secondary role (Bethlehem 1994). In the 1960s, however, the industry began to accelerate tree cultivation.

At present the country has almost 1.5 million hectares of plantations almost equally divided between softwoods and hardwoods. Some 73.5 per cent of these are private property, of which 58 per cent belong to forestry companies. The three largest companies (Mondi; Sappi; and Hunt, Leuchars & Hepburn) own 35 per cent, 30 per cent and 20 per cent of the total area in private hands.

The state owns the remaining 26.5 per cent of the planted area. This is covered mainly by pine (the government holds 44 per cent of total pine plantations) with a lower percentage of eucalyptus and other hardwoods. Private and state-owned plantations alike are concentrated in Transvaal (49 per cent), Natal (40.6 per cent) and the Cape (10.4 per cent) (Bethlehem 1994).

In the last few years, more hardwoods (particularly eucalyptus) have been planted than pines, in line with worldwide trends in market pulp (see Table 10.1). The percentage of eucalyptus in plantations climbed from 31 to 40 per cent between 1982 and 1992. The total area of eucalyptus as of 1989–90 was 538,000 hectares, mainly *E. grandis*. Annual yields in 1986–7 were estimated to be 19.3 cubic metres per hectare for *E. grandis* and 11.9 cubic metres per hectare for other species (Pandey 1992).

TABLE 10.1
New plantations in South Africa (ha)

Year	softwoods	hardwoods
1973–74	15,265	7,098
1974–75	25,168	13,770
1975–76	22,350	8,954
1976–77	15,223	6,845
1977–78	10,078	4,154
1978–79	1,095	4,790
1979–80	7,987	5,721
1980–81	8,585	1,581
1981–82	8,051	1,983
1982–83	5,745	2,516
1983–84	6,544	2,434
1984–85	10,489	4,210
1985–86	6,294	10,310
1986–87	4,584	17,308
1987–88	5,511	17,759
1988–89	7,212	22,845

Source: van der Zel 1990.

South African plantations are managed according to the final destiny of the timber. Some 38 per cent produce pulpwood, 35 per cent sawlogs and 22 per cent mining timber. However, given that there is a dwindling market for mining timber, a large proportion of these plantations may be turned over to wood chip production for export (Bethlehem 1994). Assuming that most plantations supplying the sawlog industry are in state hands, we can conclude that the majority of private industrial plantations are aimed at the pulpwood market, both national and foreign, and that it is largely private plantations which have converted the country into a significant pulp fibre exporter (Shell/WWF 1993).

During the 1980s, South Africa became a net exporter of forest products. Whereas in 1981 exports were worth 300 million rand and imports 400 million rand, in 1988 the figures were R1.5 billion and R650 million respectively (Edwards 1990). In that year forestry production came fourth in the list of national non-mineral exports (University of Natal 1990). In 1991, 56 per cent of total roundwood sold was pulpwood, and pulp and paper exports made up 40 per cent of the total demand for pulpwood (Bethlehem 1994).

The export of wood chips to Asia, especially Japan, Taiwan and South Korea, has grown markedly since it began in 1970. In 1993 1.2 million tons of chips were exported, over 97 per cent to Japan, and if the target export total of two million tonnes is reached by the end of the decade, the country will become one of the top three world chip shippers (Bethlehem 1994, FAO 1994). In 1993, the country also exported 340,000 to 550,000 tonnes of pulp (FAO 1994, *PPI* 8.1995), overwhelmingly to the North, with Japan, Austria, the US and Italy being leading importers.

The South African industry publicizes itself as having established itself through its own initiative and resources, with little or no state support (Edwards 1990). In truth, the state has not only played a pioneering role in establishing plantations and developing sawlog industries, but has also, for many decades, 'favoured the commercial forestry sector by assisting it with land use planning, training extension officers and providing financial support as well as research and development services' (Fakir 1994). The state has also taken the first steps to privatize the remainder of the plantation sector (Bethlehem 1994). Making the most of this subsidized foundation, the industry is now expanding.

The two top South African companies, Sappi and Mondi, are also moving overseas. Sappi has bought Germany's Hannover Papier and the US's S. D. Warren (the world's leading coated lignin-free paper manufacturer), as well as five paper mills in the UK. It also holds 49 per cent of the shares of the Usutu Pulpwood Company in Swaziland, whose pine plantations and kraft pulp mill cover more than 50,000 hectares (the UK's Commonwealth Development Corporation and the Swazi government control the rest) (Beer 1995, Shell/WWF 1993). Sappi has also entered into a joint venture with Mozambique's government and two local companies to develop a major eucalyptus plantation across the border (Pollett and Mander 1995). Mondi, meanwhile, holds interests in the UK and Portugal (*PPI* 7.1995).

A lack of firewood in a sea of trees

In an astonishing paradox, South Africa's supposedly 'successful' model of plantation development nevertheless fails to meet the needs of the third of the nation's people who depend on firewood as their main source of energy. It is estimated that around 12 million rural dwellers use more than seven million cubic metres of wood per year (University of Natal 1990). In the face of increasingly serious shortages, rural people are being forced to collect firewood from native woods and scrublands, degrading what scarce patches of forest remain in the country.

Even communities hemmed in by enormous plantations often do not have enough wood to meet basic energy needs. Company trees are protected by forest police, and while local people may be able to scrounge a few fallen branches, even this is illegal. In one community in Natal where Mondi planted tree farms in 1990, women and older daughters are forced, every two to three days, to walk kilometres across the valley and up hills to neighbouring farms to collect wood which they carry back to their homes balanced on their heads. 'Some of the hills are so steep my knees shiver as I climb them' said one woman; 'and on the way down, I feel my bones clicking with tiredness.' 'There is no wood left for cooking,' another villager stated; 'the tree people burnt our forests and now we walk very far to fetch wood for fires. But it's running out.' Added a third: 'It's a cold day today, but you see we have no fire. There just isn't enough wood any more' (Hornby 1994a). As Lael Bethlehem (1994), observes, 'This situation is unacceptable, both in terms of rural development and environmental protection. There is an urgent need for the development of sustainable forms of forestry which are able to provide fuel to rural people.'

More power to the powerful

Large forestry businesses have brought about substantial changes in power relations in rural areas and in control over land and other natural resources. Historically, 'commercial forestry has often been to the detriment of black people in rural areas, as plantations have generally involved forced removals' (ANC 1994). Today, vast tracts of farm holdings are reportedly being bought up by large timber firms for planting to commercial tree species at prices which are sometimes nearly double the market value of the land (Schulze 1990).

Such purchases are apparently motivated by corporations' plans to achieve control over scarce land, and to assure themselves of a constant supply of raw material (Dobson 1990). These plans have been made more urgent by the fall of *apartheid*. In the new South Africa, the previously repressed majority's claims on land may result in less land being available for plantations, so big companies are rushing to secure holdings and to plant them with trees as quickly as they can. This policy is pursued not only by the forestry industry but also by a growing number of landlords who by this means hope to increase the value of their land and thus make it more difficult for the government to expropriate it, since the constitution specifies that expropriations must be justly compensated (Fakir 1994). This, of course, reduces the amount of land the state can redistribute under its land reform programme and thus has a direct negative impact on the nation's rural poor.

Jobs, work and migration

Although figures vary, in general it can be stated that tree plantations generate less employment per unit of land than agriculture. For example, according to Fourie's (1990) surveys, agriculture employs one worker per 20–25 hectares of land, while the forestry sector requires only one for every 35 hectares. Sugar cane producers, meanwhile (although their industry is also a focus of social and environmental controversy), claim to employ one worker every five hectares, while insisting that forestry provides only one job per 18 hectares of plantation, and accuse large forestry businesses of creating 'socioeconomic deserts' (Hudson 1990).

The advance of the wood business also affects welfare and community life. For every 150–200 hectares dedicated to sugar cane, for example, there exists, according to the Cane Growers' Association, a family farm employing some 40 people, providing them with housing, schooling, medical care and retirement benefits. Such units form the basis of coherent functioning rural communities, as they support commerce, villages and towns, educational organizations, security, medical care and other social services. Forestry companies, on the other hand, consolidate a number of estates under one manager, reducing employment opportunities by 70 per cent. This leads to migration to cities, which in turn hastens the disintegration of rural communities and reduces the number and quality of services offered to those who remain (Hudson 1990, Fourie 1990).

Subsistence farmers, too, experience a decline in the quality of life following the planting of large forest monocrops. In the most common scenario, people living on a farm offer their labour to a white landowner in exchange for the right to plant their own crops and keep livestock. Even though many such farmers are descendants of people who had occupied the land before the colonizers arrived, they have no legal title to it. When landowners sell out to forestry companies, such farmers' circumstances change dramatically. The companies buy the land to plant trees and, in the face of strong resistance, try to oust families who have lived there for generations. In some cases the firms and the local people come to an agreement, but dispossession is still one of the most difficult social problems created by forestry development (Hornby 1994b). Land disputes are likely to be exacerbated when dispossessed victims of *apartheid* attempt to reoccupy land now covered by timber plantations (Bethlehem 1994).

Working conditions on plantations are far from adequate. First, the sector is marked by a low level of unionization, and many businesses still refuse to deal with worker organizations. Most plantations are 'located in conservative rural areas, where a racist and authoritarian culture often prevails. On state forests in particular, a coercive management style is common' (Zikalala 1992, cited in Bethlehem 1994). Second, plantations are dispersed widely across the country, creating great organizational difficulties for existing unions.

Large companies, moreover, have lately reduced permanent staff and encouraged the development of small subcontracting businesses to carry out tasks from planting to harvesting. This is claimed to be the result of a 'desire to see the establishment of a strong and viable informal sector serving the industry, and to move away from the concept of "paternalism" towards labour' (Edwards 1990). In reality, however, large firms are trying to lower labour costs in an industry in which levels of mechanization are very low and in which 45–50 per cent of the cost of wood production prior to transportation consists of wages. It is estimated that around 30 per cent of forestry tasks are currently carried out by subcontractors, with this figure rising to 60 per cent at harvest time (Bethlehem 1994).

In addition to drawing wages far lower than those of regular employees, contract workers receive fewer housing, education and pension benefits. In one case, workers were dismissed by a company and hired by subcontractors at 33 per cent lower pay for 20 per cent longer hours. Elsewhere, salary cuts of up to 75 per cent have been reported. Contract work is also more seasonal and less secure than standard employment (Bethlehem 1994).

Yet even for permanent employees of the large plantation companies, working conditions and levels of pay — with a few exceptions — leave much to be desired. Salaries are low, with an average working day stretching to nine strenuous hours. Companies provide workers with housing in either family units or single-sex accommodation, where conditions have been described as 'appalling' (Zikalala 1992, cited in Bethlehem 1994). Women are given the worst-paid tasks, many of which — such as bark stripping, planting and weeding — are extremely wearing and are associated with a wide variety of health problems (Bethlehem 1994). Spraying paraquat — a notorious herbicide manufactured by the UK-based firm Zeneca Agrochemicals and banned in many countries — is regarded as 'lighter work' and is mostly left to younger women employees (EJNF 1995).

Some observers expect the forest industry to use 'agri-villages' financed by the government's new land reform programme as subsidized dumping grounds for redundant labour. Thus, just as under *apartheid*, bantustans absorbed surplus labour from agriculture as it became more commercialized and capitalized, so agri-villages may become rural slums partly created by the growth of the plantation sector (Hallowes 1995).

Changes in rural lifestyles

Large-scale planting of pulpwood trees in South Africa has also had other far-reaching effects on rural livelihood and lifestyle:

* Pastoralists have been deprived of vast grasslands, including the Kwazulu area (University of Natal 1990).

* It has become harder for farmers, too, to raise livestock for meat and milk or for sale, and to thatch houses, following the conversion of grasslands into plantations.

* Reeds needed for making mats or cords used for roofing or trays have disappeared after plantations have caused small watercourses to dry up.

* When workers are expelled from farms, they can no longer plant crops or raise chickens, and poverty and malnutrition increases. 'We cry because our children have no clothes and no shoes,' said one villager. '[L]ife has been difficult since the trees came' (Hornby 1994a).

- People have been forced against their will to live in a new and threatening environment. In the most forested zones, nearly all roads pass through plantations at some point. This unfamiliar landscape inspires fear of 'wild animals' such as snakes, wild pigs, lions and other unknowns (Hornby 1994a).

- Fire has increasingly threatened the lives and livelihoods of rural inhabitants (Dobson 1990).

Water: a scarce and disputed resource

While the director of South Africa's Forest Owners Association, like his counterparts elsewhere, has declared that he does not wish to become embroiled in the 'emotive' issues of environmental conservation (Edwards 1990), the effects plantations have on water and biodiversity are of growing concern to many South Africans. The country's large-scale tree monocrops have recently been dubbed 'biological deserts' and a 'green cancer' (Dobson 1990, Edwards 1990).

The area of greatest environmental concern has traditionally been water. In South Africa, unlike most other countries, both opponents and supporters of industrial tree plantations accept that they have a big impact on water resources. This consensus is due largely to the fact that, as a result of the relative aridity of the country, which receives an average rainfall of only 444 millimetres per year (van der Zel 1990), a great deal of scientific work has been carried out on the subject. Studies were initiated as far back as 1923, and the first catchment experiments were started in 1940 to monitor how *Pinus radiata* plantations affected hydrological systems. These experiments — and others started in 1945, 1955 and 1970 in other catchments — have gained South African catchment hydrology a world reputation.

One finding of particular importance is that, in South Africa, large-scale industrial plantations consume a greater amount of water than shorter vegetation types such as scrub, herbs and grasses (Le Roux 1990). In one case, pine trees were found to consume 1,080 millimetres of water, and grasses 850 (van der Zel 1985, cited in Le Roux 1990). In 1982, van der Zel proved that in one area of grasslands with 1,000 millimetres of annual rainfall and a runoff of 200, afforestation reduced runoff by half. A six-hectare tree plantation in this area, therefore, would reduce runoff to an extent equivalent to the water required by one hectare of irrigated wheat (Le Roux 1990).

Groundwater as well as surface water is affected. Trees tend to extract water from deeper layers within the soil than do grasses, then dispersing it through evapotranspiration (Fourie 1990). Streams and wetlands fed by grasslands can thus be endangered by plantations.

South African research on catchment hydrology under tree crops led eventually to the enactment of the 1972 Afforestation Permit System (van der Zel 1990). This system established limits on the area to be afforested in each catchment area, based on accumulated data showing how plantations had already modified the local hydrology. Each catchment and sub-catchment area was categorized, with each class being allowed a certain percentage of tree cover. In Category I, no planting is allowed. In Category II (where water scarcity problems already exist), only limited planting is allowed, and the planted area must not cause mean annual runoff to decline by more that five per cent. The rest of the country falls into Category III, where plantations must not cause annual runoff to drop by more than ten per cent (van der Zel 1990). This system seeks to spread the impacts of runoff reduction more homogeneously across several sub-catchment areas.

Yet even with this system in place, water problems remain, and many water users — particularly in agriculture — are demanding improvements. The Natal Agricultural Union, for example, is calling for modifications to the current permit system, claiming that 'large scale afforestation of river basins is having a detrimental effect on the hydrological cycle of many of Natal's rivers and it is creating hardships for riparian farmers downstream' (Fourie 1990). The director of the Wildlife Society of South Africa states that

> streams which once carried perennial water before afforestation have been converted into dry watercourses or streams with erratic flow. . . . Downstream users have been adversely affected by such programmes. Life-forms dependent upon these streams have disappeared (Cooper 1990).

Other critics point out that the 1972 data on which the Afforestation Permit System was based are no longer valid. New species are being planted, and new varieties and clones with far faster rates of growth and water use have also been developed (Bethlehem 1994). In addition, commentators have recommended that the system be more finely tuned to each individual catchment area. This would involve barring tree crops which reduce runoff by more than ten per cent not only from catchment areas as a whole, but from subdivisions within them as well (Fourie 1990). The system is also being asked to take into account the impacts of afforestation on biodiversity, scenic beauty, soils, and so on. It is claimed that in many cases the current rules are being broken because of a lack of effective controls — especially by the state's own companies (Bethlehem 1994).

Biodiversity in danger

Industrial pulpwood plantations affect biodiversity in South Africa in many ways. Localized impacts are generated by the trees themselves. Generalized effects result from plantation management and dispersal of seeds from plantation trees. Studies in Natal show that 236 bird, 97 reptile, 57 amphibian, 55 mammal and 13 fish species will be affected, to a greater or lesser degree, by future plantation development. Among these are several endangered species: 24 birds, ten mammals, six reptiles, two fishes and an amphibian (Porter 1990). None of these impacts have been taken into account in current regulations, which consider only the hydrological effects of plantations.

One of the most widespread threats plantations pose to biodiversity relates to the way they alter the relationship between fire and the natural landscape. First, fires in extensive plantation blocks are nearly impossible to control once they have started. Such fires not only kill off species which have adapted themselves to life in plantations, but also affect the organic layer of the soil, its microfauna, the infiltration and runoff of water, and so on. In attempts to prevent these fires, commercial foresters set controlled fires annually in the firebreak zones, generally located in botanically sensitive areas which are home to endangered native species (Porter 1990).

Since the law prohibits controlled burning between July and October due to the risk of fires getting out of control, preventive burning of natural vegetation is done in the wetter spring and summer months. These controlled fires have adverse effects on both native flora and native fauna, especially as they coincide with the plant growth season and the period of greatest reproductive activity among invertebrate, bird and rodent species. Such fires destroy animals' food supplies (leaves, grasses, flowers, seeds, nectar, pollen, and so on), niches where they live and breed, and cover that provides a refuge from predators (Porter 1990).

Agrochemicals used on plantations also have a severe effect on biodiversity. Included are a wide range of weed-killers such as Garlon (triclopyr, manufactured by Dow Chemical), Tordon (picrotam, also a Dow product), Gramoxone (produced by ICI) and Roundup (glyphosate from Monsanto), which are generally applied in the first three years of a plantation cycle and again at the time of final felling. Insecticides, meanwhile, are used to combat wattle bag worm, pine woolly aphid, and leaf-cutting ants, which attack the acacias and pines. Rounding out the cocktail of industrial poisons used on pulpwood plantations are rodent control products such as Storm, Finale and Rattex. Plantation managers often also hunt grey duikers and bushbucks and set traps for porcupines and Samango monkeys, which are an endangered species (Porter 1990).

Apart from the above impacts, large industrial plantations affect biodiversity in grassland, wetland, and native forest ecosystems.

Grasslands

Grasslands support a wide variety of native flora and constitute the sole habitat of numerous animal species. They are home to the blue swallow, the most endangered national bird species, whose breeding areas coincide with many of the main plantation zones. Other species such as the Oribi antelope, the Stanley bustard, the blue crane, the bald ibis and many other threatened animals also depend on natural grasslands which are disappearing under plantations (Cooper 1990).

Plantations also affect a great number of other less charismatic species, which are nonetheless just as important in grassland ecosystems. Grasshoppers, for instance, are important as indicator species, as they have very specific habitat requirements. Studies in Natal revealed that exotic pine plantations had effects on communities of 26 regional grasshopper species, even in areas relatively far from the treeline (Sanways and Moore 1991, cited in Barnett and Juniper 1992).

Several grassland plant species are also in danger of disappearing, including some used locally for straw roofing and medicine. Because pulpwood plantations have been established without necessary baseline studies having been done, many species may have already become extinct without anyone knowing (Bainbridge 1990).

Wetlands

By reducing runoff and groundwater supplies through water consumption higher than that of the natural grasslands, plantations also indirectly affect wetlands. Especially in the dry season or in periods of prolonged drought, species which depend on an aquatic medium for their survival can be deprived of water and disappear (Porter 1990, Cooper 1990).

Among the species affected is the wattled crane, which requires large areas of wetland to nest and extensive neighbouring grasslands for the raising of its young. The plantations — which simultaneously take over the grasslands and dessicate wetlands — have been responsible for the disappearance of several nesting sites of this species, increasing the threat of its extinction. Other aquatic flora and fauna are affected when debris from felling washes into watercourses, releasing toxic organic compounds like tannin. In other cases, felled trees and waste from felling block watercourses, driving out species which cannot adapt (Porter 1990).

Native forests

South Africa's scarce existing indigenous forests are being damaged by pulpwood plantations in several ways. Most importantly, exotic trees are being planted right up to the margins of native forests. This results in the destruction of the ecotone (the area where grasslands and forests meet), which is an important habitat for both woodland and grassland fauna. Plantation trees may also shade out trees on the edge of native forests (Cooper 1990), and when commercial plantations adjoining forests are logged, many felled trees fall onto the forest, causing extensive harm. Finally, plantations often result in non-plantation areas being invaded by exotic species. Patula pine, black wattle and accompanying species, for example, are capable of replacing entire native ecosystems (Bainbridge 1990, Porter 1990).

Soil degradation

The impact of plantations on soils has thus far received little attention in South Africa. Current legislation appears to be based on the assumption that plantations improve the soil: unlike planters of agricultural crops, planters of forest crops are exempted from having to obtain permission from the appropriate state department before ploughing previously uncultivated land (Le Roux 1990). Nevertheless, notable soil losses have been observed during the period between initial planting and canopy formation. These losses have become more serious over recent years as planters have begun to act on studies which conclude that the more intensively soils are prepared prior to planting, the faster trees will grow. The resulting increase in erosion has eventually led to the promulgation of conservation rules for the planting period. However, these guidelines still fall far short of dealing with the whole problem and apply only to slopes of over 20 per cent (Le Roux 1990).

Erosion is also caused by roads within plantations and by extraction methods. As a result, soil is lost both when a plantation is established and when it is harvested. To this is added the damage caused by disking and clean cultivation of even the most fragile soils in firebreaks — a practice which is required by insurance companies (Le Roux 1990, Porter 1990).

Conclusions

South Africa offers an excellent example of a type of forest development which prioritizes the interests of the powerful without taking into account the needs of the majority, meanwhile degrading the environment. Large sectors of the South African population need land, firewood and water to survive. What does afforestation offer which can satisfy these basic needs? Large timber monocrops occupy vast tracts of land and expel the rural population, who thereby lose access to the resources they subsist on. Forced removals and rural–urban migration are one consequence.

The example of firewood shows perhaps more clearly than any other that forest development policy in South Africa does not even consider satisfying the needs of the majority. As noted above, a third of the South African population use wood as their primary source of energy. Common sense would seem to indicate that the satisfaction of this need should be the key element of any tree-planting policy. However, after years of intense afforestation, with 1.5 million hectares of land now given over to plantations, people have less wood available than before the process began.

For a country which must concern itself with water scarcity, it would also seem that water conservation should be an absolute priority. While the state has in fact taken hydrology into consideration, timber plantations, which generate such little value per hectare while consuming such vast quantities of water, do not seem, in the eyes of many South African observers, the most intelligent choice from a national point of view. Even if it is assumed that monospecific crops of exotic trees are necessary, there are still two possible options: large scale afforestation or the incorporation of tree planting into farming. Unfortunately, it is the first — which generates worse social, economic and environmental damage — which is still predominant.

However, there is still hope. The African National Congress's forestry policy clearly states that:

> The potential benefits in terms of fuelwood supplies, watershed management and the prevention of soil erosion, as well as other industries tree-planting could stimulate, have been overlooked. Forestry research has failed to focus on tree varieties which could be used for the purpose of social forestry . . . The ANC believes forestry can play an essential role in rural development through social forestry programmes. The design and implementation of these would be devised in collaboration with rural communities, and would seek to satisfy social, economic and environmental needs (ANC 1994).

Chapter 11

Indonesia: Deforestation and Repression

Early in 1994, the countryside around what was formerly the little village of Kerinci, 10 kilometres south of Pakanbaru, Sumatra, was witness to an extraordinary scene. Working to a plan formulated by Helsinki-based consultants Jaakko Pöyry, some 4,000 Indonesians laboured day and night, often in mud and rain, to build the biggest single-line pulp mill in the world. Under the eye of a Finnish manager, an immense soda boiler from Finland's Tampella firm was assembled on the three-square-kilometre site, supplemented by complicated machinery from Finland's Kone, Valmet, Ahlstrom, Sunds Rauma and Outukumpu; Sweden's Sunds Defibrator, Noss and Asea Brown Boveri Fläkt; Japan's Mitsubishi Heavy Industries and Nippon Sanso; Canada's Chemetics and Bailey; the US's Cranston and Solarturbines; Germany's Siemens and Voith; Britain's ICI; Taiwan's Teco; and India's Ion Exchange (*Helsingen Sanomat* 23.1.1994, *PPI* 9.1994). An integrated 280,000 tonne-per-year paper mill was meanwhile also in the works.

Built to convert four million cubic metres of wood into 750,000 tonnes of pulp each year, the Riau Andalan mill will cost its owner, the Raja Garuda Mas firm, US$750 million. Two-thirds of its output is slated for export. To satisfy the mill's appetite for wood, the equivalent of one log truck must pass through its gates every three minutes from 160,000 hectares of logged-over timber estates southwest of the mill and sources elsewhere. Some 4,750 cubic metres of water per hour will meanwhile course through the plant from the nearby Kampar River. Requiring a specially-built port and 45 kilometres of railway to move wood and wood products, the mill will employ a mere 1,000 people — $750,000 of capital investment per mill job. (Wright 1994; *FT* 14.12.1992; *PPI* 7.1994, 9.1994, 7.1995; *DTE* [London] 4.1994).

Concentration

Riau Andalan, as one of a number of enormous pulp mills which have been springing up across Indonesia's 'hinterland' island of Sumatra, is in many ways a fitting symbol of a boom which has seen the country's pulp production rise from 167,000 tonnes in 1983 to 1.4 million tonnes in 1994 and paper production increase from 377,000 tonnes to 3.05 million tonnes over the same period (*PPI* 7.1995, 8.1985; Wright 1994).

For one thing, the mill exemplifies a trend toward concentration of both production and ownership. As average Indonesian pulp plant capacity has leapt from around 5,000 tonnes per year in 1970 to 85,000 in 1991 and 217,000 in 1994, the explosively-growing industry has gravitated ever more clearly into the hands of Indonesia's biggest business families, many of whom have built their fortunes through commercial logging or plywood. Reinforcing this trend is a recent government proclamation requiring that any pulp and paper firm exploiting an industrial timber estate (Hutan Tanaman Industri or HTI) must develop an associated site to bring in Javanese labourers under the transmigration programme; only the largest companies are likely to have enough capital to develop such sites (Brooks 1994).

All of the largest new pulp installations are associated with some of Indonesia's most influential business families:

- Riau Andalan's parent conglomerate, Raja Garuda Mas (RGM) — which also runs the 240,000 tonne-per-year Indorayon mill in northern Sumatra and is building a 750,000 tonne-per-year pulp mill on a 300,000-hectare site in Sarawak, Malaysia — is headed by Harvard-educated Sukanto Tanoto (Tan Kaung Ho), the 'timber king' of northern Sumatra (*NST* 16.3.96). Among the subsidies RGM has attracted is a US$500 million loan for Indorayon from the Domestic Lending Board.

- The 450,000 tonne-per-year Tanjung Enim Lestari (TEL) mill in Musi, south Sumatra is a joint venture of Pangestu Prayogo's Barito Pacific Timber Group (Indonesia's largest wood industries group and one of the world's biggest plywood makers, holding over 4.5 million hectares of Indonesia's forest land and a one-million-hectare concession in Viet Nam); President Suharto's daughter Siti Hardiyanti Rukmana; and a consortium of Japan's Marubeni, Nippon Paper and the Overseas Economic Cooperation Fund. Starting up in 1997, the mill has a raw material catchment area of at least 300,000 hectares.

- Two giant pulp mills in Perawang, Sumatra, with a combined capacity of 790,000 tonnes per year, are run by Indah Kiat, one of around 135 subsidiaries of the Sinar Mas group. In addition to producing one-third of Indonesia's paper output (the firm is Southeast Asia's largest producer of pulp and paper and the largest stationery producer in the world), Sinar Mas also shares the world's largest holdings of palm-oil lands as well as interests in chemicals, finance, banking, hotels, telecommunications, pig-breeding, and, in collaboration with Suharto's children, various real-estate ventures. Sinar Mas is directed by Eka Tjipta Widjaja (Oei Ek Tjhong), and five of his eight children hold top posts in Sinar Mas companies. Sinar Mas also boasts another huge pulp mill called Wirya Karya Sakti at Jambi, Sumatra, with a rated capacity of 410,000 tonnes per year. Feeding the Sinar Mas mills are concessions totalling over 540,000 hectares (Soetikno 1993, *Paper* 4.2.1992, *Paper Asia* 8.1991; *PPI* 2.1994, 7.1995; *TAPPI Journal* 6.1995).

• Mohammad ('Bob') Hasan (The Kian Seng), Indonesia's top timber businessman, chair of the loggers' and wood-processing trade association, and longtime Suharto crony, has also moved into pulp and paper. Hasan holds interests in the 165,000 tonne-per-year Kertas Kraft Aceh cement-sack mill in northern Sumatra — built as a joint venture with the US's Georgia-Pacific and partly owned by the state and by Suharto's son Sigit Harjojudanto — and in a Korean/Hong Kong joint venture called Aspex Paper. In 1997, Hasan is also planning to start up a 472,000 tonne-per-year pulp operation in east Kalimantan called Kiani Kertas, whose industrial plant alone will cost around $700 million.

Export focus

Although Indonesia's pulp and paper industry has high hopes of building a bigger domestic market among the country's 190 million people (per capita consumption, now only 10 kilogrammes per annum and concentrated 60 per cent in Jakarta, could rise quickly), new capacity for both pulp and paper of the type which Riau Andalan represents is focused largely on exports.

In net terms, the country became self-sufficient in paper in 1987, with paper exports (mainly fine paper and packaging boards) rising from 200,000 tonnes in 1990 to 600,000 tonnes in 1994. (In 1993, Indonesia also imported over 120,000 tonnes, or US$168 million worth, of paper and board — a third from the Nordic countries, another third from North America, and much of the remainder from Japan and Germany. Sinar Mas, for example, has been buying pulp and paper on the open market to supply domestic customers, even while continuing to supply overseas customers [UN 1994, Sonnenfeld 1995].) Pulp exports, in 1994 around 300,000 tonnes, are expected to triple by 1996, and Indonesia may soon be shipping 60 per cent of its total production abroad (Wright 1994; *PPI* 2.1994, 7.1994). Part of this market is being created by Indonesia's own Sinar Mas conglomerate, as it invests in paper plants in Bombay and in Ningpo and Cheng Chiang, China which can take Indonesian pulp as raw material; and Raja Garuda Mas, which is also planning two paper mills in China. Sinar Mas already exports nearly half of its production of pulp and paper, while Indorayon's north Sumatra plant has been exporting 60 per cent of its pulp output to Japan, Southeast Asia, Europe, Taiwan, the Middle East, China and Korea (*PPI* 8.1993, 2. 1994, 7.1994, 8.1994, 10.1994, 5.1995; WALHI 1992). Indonesia as a whole has recently risen to be the 17th largest paper producer and the 18th largest pulp producer in the world (*PPI* 7.1995).

Just as Indonesia shifted years ago from exporting raw logs to exporting value-added processed wood, it is now shipping far more pulp and paper than wood chips. In 1993, tens of millions of US dollars' worth of hardwood chips left the country for Japan and elswhere in east Asia, but the trade in paper and pulp is already valued in the hundreds of millions. (Indonesian hardwood chip export figures differ from Japanese import figures by a factor of three, making it difficult to quantify the trade precisely [UN 1994, FAO 1994].) In 1993, 44 per cent of Indonesia's pulp exports went to South Korea, Japan and Thailand, and the rest to the Philippines, Bangladesh, Malaysia, Pakistan, India, Viet Nam, Singapore, China and other Asian destinations (UN 1994). These markets, together with Taiwan, are likely to remain at the centre of the Indonesian industry's overseas marketing strategy as the country becomes a net pulp exporter.

Indonesia's foreign paper and board shipments meanwhile, go to a somewhat different range of countries. In 1993 half of the country's exports ended up in Hong Kong, Malaysia, Singapore and South Korea, with the bulk of the remainder going to Australia, China, Thailand, Egypt, Sri Lanka, Saudi Arabia, and the US. The paper sector's growing export orientation helps further to concentrate industry ownership. As Sinar Mas executive Suresh Kilam explains, '[t]he companies which cannot export will go under, with the smallest mills being closed down rather than taken over. With an annual capacity of only a couple of thousand tonnes, they are not economical' (*PPI* 2.1994).

Foreign involvement

A third industry characteristic exemplified by Riau Andalan is heavy Northern involvement. Most of the hundreds of millions of dollars spent to build and plan the wood supplies for new mills goes to Nordic, Japanese, and North American suppliers and consultants and their consortia (Allen 1992; *PPI* 8.1993, 2.1994, 10.1994; DTE 1991; *Paper* 4.2.1992). Sinar Mas's Perawang and Jambi mills, for example, carrying total price tags of US$520 million and $630 million, use pulp lines supplied by Sweden's Kværner Pulping, power and recovery boilers purchased from Ahlstrom, a Kone woodyard, and other equipment from Andritz Sprout-Bauer and Asea Brown Boveri Fläkt (Wright 1994, Pappens 1993, *TAPPI Journal* 6.1995). Finland's Rauma Repola, meanwhile, provided washing and bleaching equipment for Indorayon's pulp and rayon project in Porsea (DTE 1991), with the mill's waste-based energy turbine being supplied by Ahlstrom and Sunds Defibrator of Sweden and its wood yard by Elof Hansson (*Paper* 4.2.1992). Bob Hasan's Kiani Kertas firm in east Kalimantan has meanwhile ordered a woodyard and 472,000 tonne-per-year fibre line from the US's Beloit and other machinery from IMPCO (Ingersoll-Rand). Also important in Indonesia's pulp and paper industry are German, Austrian and Swiss technology and engineering services suppliers such as Voest-Alpine, Voith Sulzer, Klockner Stadler Hurter, Bielomatik and E.C.H. Will, as well as Sweden's Asea Brown Boveri and Belgium's Novarode (*PPI* 2.1994; Sonnenfeld 1995).

Japanese papermakers have also long been interested in Indonesian wood fibre. Kojin, Kanzaka Paper, MDI, and what were then Sanyo-Kokusaku Pulp and Jujo Paper were all involved in the Chipdeco mangrove chipping plant in east Kalimantan, and Marubeni in the chipping of Southeast Asia's largest mangrove forest at Bintuni Bay, West Papua. Barito Pacific Timber's 450,000 tonne-per-year pulp mill, meanwhile, is to be one-third owned by Japan Sumatra Pulp, an investment company jointly owned by Marubeni, OECF, and Nippon Paper, who together put up US$100 million for the project. Japanese pulp and paper equipment is now also being sold to Indonesia, aided by the attractive financing made possible by Japanese trading houses. Indah Kiat's Perawang plant, for instance, uses machinery from Hokuetsu Paper Mills of Japan as well as from Beloit-Mitsubishi, and Kotobuki, together with Valmet, supplies Tjiwi Kimia, Sinar Mas's huge paper firm (*PPI* 8.1993, 2.1993, *Paper* 4.2.1992).

In order to land such machinery or forestry contracts or negotiate partnerships, industrialized-country firms have to engage in a great deal of stage-setting beforehand. Here a local presence is crucial, and inter-industry collaboration extremely useful. Sweden's Sunds Defibrator and Ahlstrom Machinery, for example, have regional offices, and work together with each other and with Kværner and Valmet to provide complementary equipment. Valmet sells its machinery through Swedish-based CellMark, one of the largest commodity traders in the world, which plans also to join Marubeni in marketing pulp from Barito Pacific's TEL mill (see below) (Kuroda 1995).

Acting as brokers, planners and coordinators are forest industry consultancy companies such as Finland's Jaakko Pöyry and Canada's

Sandwell and H. A. Simons. Such consultants profit from adjusting resources to technology, bringing together Northern machinery suppliers, local business groups, and Indonesian leaders with power over land, forests, finance and labour in a way which benefits all. Working through a network of old colleagues, friends and like-minded technocrats in overworked donor agencies who are happy to be relieved of onerous planning and monitoring duties, the consultants have been able to use public funds to build up a private forestry industry sector in Indonesia which ensures them a continuing stream of future contracts. For example, a 1984 contract with the World Bank and the Indonesian government to do a sector analysis of the country's paper and pulp industry helped Jaakko Pöyry land over 30 subsequent contracts to plan or implement public and private sector projects to supply mills with pulpwood from natural forests or plantations. In addition, the company has picked up scores of contracts — some of them subsidized by Finnish taxpayers through Finnish Export Credit and the bilateral 'aid' agency FINNIDA — to plan or engineer pulp or paper mills for Indonesian clients or do market surveys for Western machinery manufacturers (Jaakko Pöyry n.d.). It has also benefited from other agreements, as when Finnish Export Credit and FINNIDA granted a 13-year interest-free loan worth US$4 million for a forestry development and training centre for the Indorayon firm in northern Sumatra (DTE 1991, ODA 1992, Westoby 1987).

Using Canadian government lobbying and handouts, H. A. Simons and Sandwell also mediate politically between the interests of Northern and Indonesian elites. In March 1992, for instance, the Canadian Embassy in Jakarta hosted a mission of 35 companies from the pulp and paper sector to promote Canadian pulp and paper expertise and technology through seminars and top-level meetings with government and trade officials and local industrial officials and associations. CIDA, the Canadian bilateral 'aid' agency, is meanwhile funding a feasibility study for a plantation and pulp and paper mill in Sorong, West Papua owned by the Kayu Lapis Indonesia group. While Sandwell provided engineering, design, training and operation services to Indorayon, and also helped to plan Barito Pacific's TEL mill (DTE 1991, *Paper* 4.2.1992), H. A. Simons is supervising site preparations for Bob Hasan's Kiani Kertas plant in east Kalimantan (*PPI* 9.1994).

Foreign investors and financiers are also joining Indonesia's pulp and paper tycoons in helping to get the industry's immensely capital-intensive projects off the ground. By 1990 the pulp and paper sector was already Indonesia's seventh largest in terms of foreign investment, receiving $730 million of the total $8.75 billion invested that year (WALHI 1992).

Helping to start up Sinar Mas's lucrative exploitation of natural forest, for example, was the Taiwanese firm Chung Hwa Pulp, which shipped one of its used pulp lines to central Sumatra in 1984 as part of its joint venture with Sinar Mas subsidiary Indah Kiat (DTE 1991, *PPI* 8.1993); today Chung Hwa and Yuen Foong Yu own a quarter of Indah Kiat's shares. Similarly, the partly state-owned Finnish paper giant Enso has been involved in forestry in Indonesia since 1981, most recently as a consultant for a FINNIDA-funded plantation trial and 'reforestation' project. Enso is now teaming up with state-owned forestry company Inhutani III and Gudang Garam (Indonesia's biggest cigarette firm) in a ten-year plan to develop an approximately 139,000-hectare acacia plantation in Sangau, west Kalimantan, partly on grasslands. Some 80,000 hectares of the land is already claimed by local residents (*JP* 22.7.1994; *PPI* 8.1994; *Enso Vision* 9.1994; Junus 1994; Rasmusson 1995). Enso and its state backer, Finland's Fund for Industrial Cooperation, are contributing $30 million of the total $100 million cost. Enso may also take a stake in a 500,000 tonne-per-year pulp mill which would be built on site after the acacia matures, and is to set up a liquid and food packaging board plant with Gudung Garam, which it would supply with chemical pulp and knowhow (*PPI* 8.1994).

Foreign finance is also crucial to industry development. Some 70 per cent of the capital needed for the US$1 billion Barito Pacific TEL project, for example, is being sought through a flotation of an international stock issue with the assistance of Morgan Grenfell Asia. Finance will be in the form of loans for which sales of the product will serve as security (*JP* 9.4.1994). Sinar Mas, in the guise of a holding company called Asia Pulp & Paper, has filed an initial public offering worth $390 million in the US to help finance mills in China, India and Indonesia. And Raja Garuda Mas has set up a Bermuda-based holding company, Asia Pacific Resources International Holdings, which recently carried out a $150 million offering in the US and Canada to obtain finance for projected operations in both Indonesia and China. The US's Chemical Bank, as well as other Northern banks, are helping to organize similar offerings for other mills in Indonesia (*PPI* 8.1993, 10.1994, 5.1995; Kuroda 1995; DTE 1991; Sonnenfeld 1995). Foreign finance is needed not only to build big mills, but also to set up huge plantations, which, at a cost of at least $3,000 per hectare, require injections of capital 'beyond reach' of domestic groups (Data Consult 1990).

International promotion of the mainstream pulp and paper industry in Indonesia, as elsewhere in the South, involves a reapportioning of risk from the private to the public sector, and from North to South. The inequities involved are obvious. The immense price tags of Kværner or Tampella machinery constitute a far greater proportion of Gross National Product in Indonesia than in the US, straining not only clients but also public guarantors. And while the hard-won knowledge of a subsidized Jaakko Pöyry consultant about how to work Indonesian bureaucracies is a permanent gain for the firm in its future dealings abroad, Indonesian magnates' profits in a volatile global pulp and paper market may be less secure. Still more precarious are the livelihoods of the ordinary people from whose land, water and labour officials and pulp and paper firms must squeeze margins and taxes.

It is often remarked how debt and poor terms of trade provide incentives for Indonesia to rush to cash in on its forest land. As income from oil and gas, Indonesia's top earners, falls, the contribution of the forestry sector to GNP, foreign exchange earnings and the payment of the national debt becomes more significant. After log exports were banned in order to add value to forest exports, and after the huge wood-processing industry began to face the prospect of eventual raw materials shortages from existing concessions, pressures grew to focus on pulp and paper. According to Eddie E-Tak Shaw of Sinar Mas,

> [i]n 1980 the government set aggressive targets for the development of plywood and softwood production. Indonesia was the number one plywood manufacturer in the world. But ten years on the government has realized that with 60–100 years' growing time [the plywood industry] couldn't survive. . . . it initiated a study into how to develop other wood-related industries. And there was environmental pressure. Paper and pulp was the right one to go for (*Paper* 4.2.1992).

Thus the government has spoken of making Indonesia into one of the top ten paper producers in the world by 2004, with as much as ten per cent of its land under tree plantations. Industry Minister Hartarto claimed, indeed, that it would become 'the biggest paper and pulp producer in the world'. In addition to opening up vast tracts of forest land to the industry, the state has given it incentives such as exemptions on import duty for pulp and paper machines (Data Consult 1990).

Whatever may lie behind such thinking, translating it into working logging concessions, tree plantations, pulp and paper mills, and other types of forest-consuming technology would be impossible without the concrete activities of foreign consulting firms and foreign investors, as well as their helpers in their countries' bilateral 'aid' agencies, export-credit organizations, and so forth. It is largely through these activities that opportunistic tycoons and others are enabled in practice to integrate more and more of Indonesia's forest lands into the world economy (incidentally helping to keep world markets flooded with fibre products). The destructive results are sketched in the next section.

Plantations and deforestation

A fourth characteristic of the new Indonesian pulp mills is that they are fed in their initial stages, by and large, by native (though often logged-over) forests. While plantations are often advertised as improving 'degraded' or 'unused' lands, the most obvious candidates for this treatment — anthropogenic, homogeneous *alang-alang* grasslands colonized by *Imperata cylindrica* — are difficult and expensive to convert, due to low soil fertility, the grass's toughness, and their tendency to be scattered in patches over a large area. Such grasslands are also often used by local peoples (WALHI and YLBHI 1992, Sonnenfeld 1995). While the state timber firm Inhutani III and Finland's Enso have established plantations on grasslands in west Kalimantan, it generally makes far more sense for pulp tycoons with privileged access to the Ministry of Forestry, which has jurisdiction over about two-thirds of the country's surface area, to take out what are essentially large, contiguous logging concessions at a rent of approximately US$0.30 per hectare per year, clear-cut them, claim the wood as raw material for pulp, apply for HTIs, and then replant the land with pulpwood monocultures. Plantation entrepreneurs who do so are entitled to equity capital and no-interest loans from the Ministry of Forestry. Such loans cover, in effect, 20 per cent or more of the costs of setting up and maintaining HTI plantations through an eight-year cycle (Fearnside 1993). Plantation companies can also use revenues collected from logging companies and earmarked for rehabilitating logged-out concessions as a way of financing tree-planting. Logging concessionaires' ability to hold onto deforested land and convert it to tree plantations has meant that it is less available to inmigrating smallholders, who may therefore have to invade forest land elsewhere.

So easy, in fact, has it been for business to gain or keep control over forest land by claiming to be interested in pulp and paper — by 1992, 37 companies had submitted applications for seven million hectares of pulp estates, although only eleven bothered to undertake feasibility studies on estate development — that the government recently had to stop granting the concessions (*PPI* 9.1994). The state has also helped companies acquire land by discouraging enforcement of customary *adat* claims to many of the territories affected. Such claims, although they are recognized in theory, are in practice overridden by the demands of 'development'. As former Forestry Minister Hasjrul Harahap said in 1989, 'In Indonesia, the forest belongs to the state and not the people . . . they have no right to compensation'. In south Sumatra, out of ten million hectares, farmers hold only 1.5–1.7 million hectares, while 19 companies own 1.9–2.0 million (Kuroda 1995).

Thus, until at least 2002 Riau Andalan will be harvesting 50 species of native tropical hardwood from its concessions while waiting for *Acacia mangium* and eucalyptus plantations to mature on cut-over sites. By the end of 1993, however, only 4,000 hectares of clear-cut forest had been replanted (*PPI* 9.1994). Elsewhere in Sumatra, Sinar Mas's 790,000-tonne-per-year, 1.2 square-kilometre Perawang installation will consume 200 square kilometres of old-growth forest per year until the year 2000, much of it cut by transmigrant labour, before switching over to acacia; the company's 410,000 tonne-per-year Jambi mill further south will denude its 2,400-square-kilometre concession of logged-over forest at perhaps half that rate (*PPI* 10.1994). By 1988, the first of the two Perawang mills alone was reported to be consuming 400,000 tonnes of wood yearly from over 100 species of tropical hardwood. 'Basically,' company vice president Eddie E-Tak Shaw explained, 'we are looking for forest which can be clear-cut and replaced with eucalyptus and acacia' (*PPI* 1.1988). The site's new 410,000 tonne-per-year pulp mill will also feed on mixed tropical hardwoods until 1999 before switching to acacia from plantations which will ultimately total 140,000 hectares, of which 60,000 have already been planted. Sinar Mas's Wirya Karya Sakti mill in Jambi, Sumatra, also at 410,000 tonnes per year, had by 1994 planted only 7,000 hectares of pulpwood trees on its 241,000-hectare concession of logged-over secondary forest, although there are plans to add 15,000 hectares per year from 1994 (*PPI* 8.1993).

Indorayon's 240,000 tonne-per-year pulp and rayon mill in north Sumatra, which started off using 86,000 hectares of old stands of indigenous *Pinus merkusii* planted during the Dutch colonial period, has also clearcut mixed hardwood forests from a 150,000-hectare selective logging concession and planted eucalyptus there. Natural forests elsewhere are also falling prey to Indorayon, most notably 40,000 hectares of what was formerly Harionboho Protection Forest. Some 100,000 hectares of pine forests in Central Aceh, meanwhile, will feed the Kertas Kraft Aceh cement sack mill in northern Sumatra through the turn of the century, until replanted pines become available. Mills projected for Kalimantan and West Papua, such as Bob Hasan's Kiani Kertas, will also consume natural hardwoods during the first years of operation (*TAPPI Journal* 6.1995). Of the major new plants now coming on line, or about to, only Barito Pacific Timber's planned TEL mill will not start out by being mainly dependent on natural forest, benefitting from at least 300,000 hectares of concession, of which 210,000 has already been planted, mainly to acacia (Wright 1994, Allen 1992). Even here, however, natural forest has been cleared in preparation for plantations (Kuroda 1995). Although some observers expect future mills to rely in their initial stages less on mixed tropical hardwoods than on pulpwood or even palm oil plantations (Sonnenfeld 1995), industry analysts suggest that even if the government is able to develop its originally projected 4.4 million hectares of industrial plantations by 2004, pulp mills will continue to have to rely heavily on the 55.4 million hectares of designated natural forest logging concessions until well after that date (Data Consult 1990).

Legal restrictions on clearcutting or on logging on steep slopes or near watercourses, such as they are, carry little meaning in this context. The Department of Forestry itself admits that over 86 per cent of timber concession holders violate government logging rules, while the Environment Ministry confesses that 500,000 hectares of forest are cut illegally each year (*JP* 18.2.1994, *Tempo* 5.2.1994). According to the World Bank, meanwhile, 'only 2.2 per cent of cutover lands have ever had a legally required residual stand inventory, and cutting outside approved boundaries is frequent'; the Bank warns that timber companies are cutting 50 per cent more logs than can be replaced through replanting (World Bank 1994; *IHT* 22.9.1994). In 1994 it was reported that some 248 logging concessionaires had recently been found guilty of logging violations (*PPI* 9.1994). To take just one example, Indah Kiat, which, as the government acknowledges, has insufficient access to forested land to meet its pulpwood requirements legally, was fined US$1.4 million recently for employing transmigrants in its charge to carry out illegal logging (*JP* 8.9.1993, 27.9.93, 3.6.1994, 30.7.1994; *Kompas* 27.9.1993; *FT* 9.9.1994; *Forum Keadilan* 6.1.1994; DTE 1991; Zerner 1992). Companies' own conservation pledges — for example, Riau Andalan's promise not to cut endangered species or log on slopes of 23 degrees or more within 200 metres of rivers — are no more likely to be acted upon than the forestry laws themselves.

Moreover, if plantations do not mature as expected — and experience with them is scant on the generally poor soils of the Outer Islands — then pressure on native forests may increase still further (*PPI* 8.1993, 10.1994; WALHI and YLHBI 1992). So far the largest contiguous plantation of any kind in Indonesia has occupied 20,000 hectares, yet Pöyry consultants are recommending units of 30,000–50,000 hectares. The risks are considerable. Plantations of *Leucaena leucocephala* on unforested land, for example, ran into disaster during the 1980s when insects attacked on a large scale. Given that no pest-resistant species of acacia have been found, similar problems may await pulp plantations. A prominent softwood silviculture researcher has gone so far as to comment that 'Indonesia may be a disaster waiting to happen' due to hasty plantation planning (WALHI and YLBHI 1992).

No less signficant in forest destruction is the disruption of community-based resource-management systems through, for example, road-building, site clearance, land takeovers, the blocking of tributary streams and the disruption of local irrigation systems and fisheries. Indah Kiat's Riau concessions, for example, are largely on ancestral lands belonging to the Sakai people, who have been resettled outside the area. Imported labourers also play a part in deforestation. In the words of the World Bank,

> the fact that only a small fraction of [traditional] *adat* rights have been registered makes it difficult for existing forest dwellers to prevent spontaneous transmigrants from taking over traditional clan lands and, lacking the cultural traditions of the indigenous agriculturalists, farming the soils to exhaustion before moving on in search of new land (1994a).

As Charles Zerner points out in a suppressed World Bank report, the agricultural and forestry systems threatened in these various ways not only have been effective in sustaining forests in the Outer Islands, but in some cases have even enhanced biological diversity (Zerner 1992). Through displacement, impoverishment and deskilling of local people, pulpwood developments are thus undermining one of the main bulwarks of forest protection in Indonesia.

Forest fires linked to pulpwood operations are another important cause of deforestation. Logging often leaves behind large amounts of dry wood and scrub which provide ideal fuel, as does the practice of clearing scrub and remnant forest in order to plant fast-growing trees. Fires can also be deliberately set in order to provide a rationale for declaring the burned-over area a timber estate, or as an act of retaliation or resistance by evictees or underpaid transmigrant timber workers. Fires can be particularly extensive in certain areas of Kalimantan, where a layer of subsurface coal provides additional fuel: in 1982–3 a single gigantic blaze devastated 3.6 million hectares of the island's forest cover, an area 56 times the size of Singapore. Huge fires in succeeding years have wiped out further thousands of square kilometres and disrupted air traffic over wide areas. Plantations themselves, of course, often fall victim to fire, resulting in millions of dollars in damage (*Setiakawan* 1–6.1992, SKEPHI 1995).

Further subsidies

Subsidized land, wood and technology, in sum, have helped make Indonesia into one of the cheapest pulp and paper producers in the world. But the subsidies do not stop there. Salary levels are also among the world's lowest, kept that way partly through state-supplied suppression of labour movements. As the *Financial Times* wrote in 1990,

'The almost total absence of labour unrest, together with officially-sanctioned low wages in Indonesia, has been one of the main factors in attracting large amounts of foreign investment to the country'.

Near-captive labour, meanwhile, is provided by the transmigration programme for projects such as Barito Pacific's TEL pulp installation. At Sinar Mas's Perawang complex in central Sumatra, inadequately-housed transmigrants required to do illegal logging have had to wait up to three months at a time for their wages, on one occasion being paid only after they took some company vehicles hostage (*Forum Keadilan* 6.1.94). Tree-planting wages offered by the Jayanti Group to transmigrants on the company's timber reforestation site on Seram Island, Maluku, meanwhile, amount to only one rupiah per seedling planted. At a rate of one tree a minute this earns only Rp 600 per day, or about $0.30. The nucleus estate/smallholder model, which the government is hoping to adapt to pulpwood plantations, also has a record of poor wages (DTE 1991).

At a time of tightening controls in the North, lax enforcement of pollution laws (however strict they may be on paper) also attracts pulp investors. To take one example, Indah Kiat — the subject of a lawsuit threat in 1992 for its pollution of the Siak River — uses its wastewater treatment facilities when there is an inspection but otherwise only sporadically, and has consistently failed to meet deadlines for waste control. Another firm which installed over $30 million worth of expensive-to-run pollution control equipment also operates it only rarely.

During Indorayon's startup period, similarly, the company paid no attention to an official directive that it construct an aerated waste-treatment lagoon according to a government plan. When the waste lagoon it did build collapsed, sending effluent into the Asahan River, no compensation was paid to fisherfolk, who, up to 40 kilometres downstream, saw their yields fall by up to 90 per cent and were forced to find another source of income. When local NGOs found persisting high levels of pollution six months after the accident, Indorayon staff succeeded in intimidating affected villagers into not taking the case to court, appearing at meetings only in the company of district military officers, showering locals with gifts of food and offers of jobs, and insinuating that NGOs interested in the case had a hidden Communist agenda (WALHI 1992; *JP* 7.9.1992, 10.9.1992, 22.9.1992, 24.9.1992, 8.10.1992; *Suara Pembuaran* 31.8.1991, 4.9.1992; *Kompas* 17.11.1993; Anon n.d.). Following the explosion of a chlorine tank at Indorayon in November 1993, which drove away many terrified local residents, Environment Minister Sarwono Kusumaat-madja ordered the mill to halve its production pending completion of an environmental audit by an independent consultant. However, the audit, carried out by the US firm Labat Anderson, cannot do more than recommend how Indorayon can best continue to use chlorine. In November 1995, rumours of a hazardous gas leak provoked hundreds of local protesters into burning 100 houses, a radio station, and Indorayon vehicles (*FEER* 15.2.1996).

A final selling point for Indonesia in the eyes of the plantation industry is a climate which allows pulp species to mature three to four times faster than in, say, Finland, theoretically making possible shorter lead times, smaller plantations, and more efficient use of land.

Disintegration and resistance

Many ordinary Indonesians, unsurprisingly, experience the integration of local land and forests into the world pulp and paper economy as a *dis*integration of local livelihoods and relationships, and gifts of low-cost forest land and riverine waste sinks to the industry as something more closely akin to theft.

When, following independence, the Indonesian state arrogated to itself the country's forest lands in a move which has paved the way for many of today's pulpwood developments, it was not imposing its legal order on a vacuum, but on the customary and informally-enforced claims of local communities. Industrial timber estates (HTIs) and nucleus estate schemes have often functioned essentially as ways of redistributing customary lands to the wealthy — although they are advertised as ways of reclaiming critical lands, providing an alternative to logging natural forests, supplying an incentive to reforestation, and providing employment to predominantly Javanese transmigrants in a way which will prevent them from having to resort to destructive forms of agriculture on the unfamiliar soils and terrain of the Outer Islands. The effects that Zerner (1993) has enumerated for timber concessions, HTIs, and nucleus estate schemes generally also apply to pulp plantation development in particular:

> [T]he effectiveness of local resource management institutions has decreased and community control over historically owned and managed forest territories and resources has been eroded. Community nutrition has declined as the quality of hunting territories and fishing sites has decreased and access to these areas is progressively limited. Access to markets through river-based transport networks has been reduced. In addition, . . . community health and food security may have declined and the incidence of . . . sexual abuses and assaults has increased. In the vicinity of logging camps, prostitution has increased and venereal disease has spread. . . . exclusion of local communities from . . .

forest territories has often resulted in the political polarization of communities rather than their integration as citizens. Alienation, opposition and violence are among the social costs of a failure to recognize community needs and rights to equitably participate in the forest management enterprise.

In northern Sumatra, to take one example, Indorayon's clearcuts, roads and plantations have displaced thousands of Batak people in the Lake Toba area through usurping their traditional lands or degrading the environment which sustains them. Logging-related droughts have depressed rice harvests over wide areas, and a pine plantation has been established in the Sibatuloting Protection Forest, a catchment area of two river systems supplying water to 65,000 hectares of wet rice paddies and 688,000 people throughout the 'rice bowl' of northern Sumatra. Logging roads built into pine forests have destroyed rice fields and resulted in two landslides in October 1987 and November 1989 which took a total of 30 lives. Painstaking attempts to use official channels, and then road blockades, to get the company to accept responsibility have resulted mainly in arrests and military intimidation of villagers, clergy and NGOs. The company has also succeeded in dividing villagers by offering them partial compensation.

In March 1987, in addition, without permission from local elders, Indorayon began stripping ancestral pasture lands legally belonging to Batak people of the village of Sugapa and planting them to eucalyptus. A fine of 10,000 rupiah was imposed on any animal found grazing in the area where seedlings had been planted. Distraught villagers' attempts to pursue the case with local and district authorities through 1988 bore no fruit. To forestall further opposition, village- and local-level government officials tricked villagers into signing blank sheets of paper that were then turned over to Indorayon staff, who appended them to an agreement awarding the pasture land to the company. This was then shown to Sugapa residents at an official meeting at which they were protesting the abuse of their land rights. At the same time, local government officials claimed that they had already been able to induce several villagers into accepting compensation for the lost land. Enraged by these machinations, Sugapa villagers, particularly women, in whom according to matriarchal custom land rights are invested, finally exploded in April 1989, when Indorayon employees were caught sexually violating local women — an action considered by the Batak people to be the most heinous of all crimes. When the men were released by the police without charge, infuriated Sugapa women marched on Indorayon's plantation and ripped up thousands of eucalyptus seedlings planted on *adat* soil. Charged with destruction of company property, the women subsequently traveled to Jakarta to take their case to the Minister of Home Affairs. At the February 1990 trial at which ten women involved were sentenced to six months' imprisonment (later reduced on appeal to probation), they vehemently defended their rights to the property: 'The land is the only source of income that the people have. If it is planted with eucalyptus, how are we going to eat? How are we going to feed and herd our cattle?' By this time, however, Indorayon had succeeded in dividing the community sufficiently to blunt the movement for return of the land (WALHI and YBLHI 1992, Anon n.d.).

Indorayon has also provoked local resentment by blocking access to woodlands which villagers rely on for rattan, fodder or carving wood; by planting eucalyptus on an ancestral graveyard; and by demeaning villagers by forcibly overriding traditions of hereditary land transfers important to clan identity. In November 1992, farmers in Gonting Silogomon village, 156 hectares of whose land had been appropriated for an HTI to supply pulpwood to Indorayon, asked the firm's workers to stop clearing their cultivated fields. During the following days, local farmers were repeatedly threatened by military officers from the District Infantry armed with M-16s, with several shots being fired (Tjahjono 1993). Farmers from one village who agreed to grow eucalyptus for Indorayon subsequently regretted becoming involved since they were no longer allowed to pasture their animals on their land and feared Indorayon would set low prices for the wood they produced. Indorayon's pollution of the Asahan river, meanwhile, has resulted in degradation of fisheries, health problems and loss of village water supplies, while air pollution has brought on nausea and dizziness as far as 40 kilometres from the factory (*Environesia* 4–8.1990, Anon n.d.).

In south Sumatra, meanwhile, PT Musi Hutan Persada, one of Barito's timber estate management companies, when it found that the 300,000-hectare concession it had been awarded overlapped with one given to another timber company in the area, took over fertile land belonging to 200 farmers in Muara Enim regency without warning, destroying durian and jackfruit trees, rubber plantations, wet-rice fields and forest commons. Protesting farmers were arrested and their claims dismissed out of hand by the provincial governor, who claimed they were merely seeking financial gain. Minister for Forestry Djamaloeddin Soeryohadikoesoemo, for his part, claimed it was the responsibility of the company to settle the dispute — despite the fact that the concession was awarded only on the condition that fertile or privately-owned land not be used (Brooks 1994, *Tempo* 18.12.1993, SKEPHI 1995). The fate of the displaced villagers remains unclear. In other parts of Muara Enim and Muara Rawas regencies, thousands of hectares of forests and farmland have also been seized by company agents for pulpwood plantations (sometimes on the pretext of establishing oil palm or rubber plantations) without consultation with local farmers (WALHI 1995, *JP* 15.6.1995). In many cases, promised compensation for seized land in the form of rubber plantations, livestock and hospital services has not materialized (SKEPHI 1995, Kuroda 1995). Nor have villagers farming on the Tanjung Enim Lestari mill site been informed of the details of the factory (Kuroda 1995).

In Kalimantan, finally, in one of dozens of such examples, logging roads constructed parallel to a river on one plantation concession blocked tributaries, halving the river's flow and creating a malarial swamp. Filled with sediment from erosion from logging and road-building, the river became useless for transport to market or for fishing. No compensation was paid, meanwhile, for the logging of lands which had been owned and managed by the local community for 400 years. The company involved also put up a sign indicating that subsistence swidden farming on community land was a crime, and ordered local residents and Javanese transmigrants to use hoes instead. Although only the transmigrants obeyed, catastrophic erosion was the result. Another disaster was precipitated when the firm tried to 'develop' the villagers' traditional rattan cultivation methods by getting them to plant rattan in straight lines without regard for site conditions, with the result that 90 per cent of the crop died (Zerner 1993).

Elsewhere in Kalimantan, a company planned to establish an HTI, apparently for a paper mill, not on degraded land, as regulations require, but on secondary forest land. This land had been under *adat* ownership for 200 years and used, among other things, for rotational swidden and rattan gardens. Company 'experts' did not recognize these productive agroforestry systems even when walking through them. When the local community refused to allow the area to be surveyed, the firm claimed ownership of the land and promised compensation for destroyed rattan of 300 rupiah per cluster. Yet the 20-year value of returns for one cluster of rattan is estimated to lie between Rp 500,000 and Rp 1 million. That is, the compensation offered was only .06 per cent or less of the market value of the commodity in question.

Such cases underscore the role plantation projects have in degrading forests through displacing people and disrupting community-based forest stewardship systems. In contesting the actions of plantation and pulp companies — whether by appealing to local, district, provincial and national bureaucracies and parliaments, taking firms to court, blocking roads, seizing company vehicles, pulling up stakes put down by survey teams, tearing down signs criminalizing traditional agriculture, cutting company bridges, pelting company houses with rocks, or simply refusing to cooperate with firms who do not consult with residents according to locally-approved political processes — local villagers are also attempting to defend the main features in their social landscape which hold out a secure promise for the future.

In doing so, they have often worked with local legal aid groups or NGOs. On some occasions, national-level alliances have also been formed which use national law as a tool, as in the late 1980s when WALHI, a Jakarta-based NGO, began working with north Sumatra community development NGOs in surveying Indorayon's abuses with an eye to taking the firm and the government to court in Jakarta to test a 1982 statute on environmental regulation. Foreign groups have meanwhile demonstrated solidarity by asking funding agencies (for example, Japan's OECF) and foreign investors in Britain and the US to reconsider their support for projects such as Barito's TEL scheme (Tjahjono 1993, WALHI 1992, Kuroda 1995).

All alliances critical of the Indonesian pulp industry carry risks. Villager–NGO alliances may bring down charges of 'Communism' on the heads of both sides — charges which have proven fatal to so many Indonesians since the massacres of the mid-1960s. Villagers may be beaten, arrested, or coopted, while NGOs may face closure or intimidation. International campaigns, if they are seen as tied to local battles, may increase state intolerance of local groups even further.

Yet alliances can also bring gains. The lawsuit initiated by WALHI, for example, although it was lost, helped catalyse coordination among north Sumatran NGOs on broader environmental issues, and impressed again upon urban-based NGOs the importance of patient work with villagers and rural NGOs. Where popular demands are seen as coming from forums rather than from individuals or organizations, moreover, villagers and NGOs can be protected to some extent from corporate or official retribution.

As the government's investment in global public relations testifies, international campaigns can also affect the development of the pulp industry in Indonesia. A campaign involving the threat of a boycott in Europe and North America, for instance, helped force Scott Paper to withdraw in 1989 from a joint venture in West Papua which would have displaced 15,000 people. Scott's partner, Astra, Indonesia's second largest business conglomerate, has vowed to continue the scheme, but financial difficulties have slowed its progress.

Chapter 12

Thailand:
From 'Reforestation' to Contract Farming

Incentives for establishing commercial pulpwood plantations are as strong in Thailand as in Indonesia: a booming economy, strategic location, lavish subsidies, and local and foreign elites eager to invest. But competition for available land and forest, a less repressive political climate, and, most of all, heavy resistance, have so far prevented a plantation boom on the scale of that in Indonesia. Continuing pressure from the paper and pulp industry and its allies, however, has led to a seesaw battle for high stakes across large areas of the country between plantation proponents and local villagers and environmentalists.

Manufacturing and export growth

Much of the impetus for establishing new plantations stems from the Thai economy's growth rate, one of the highest in the world over the past decade. Rises in paper and paperboard demand, however, are driven not simply by aggregate economic expansion, but by manufacturing and export growth in particular. Packaging, for example, accounts for a staggering two-thirds of the paper used by the country. Revealingly, moreover, the 1994 paper and paperboard demand increase of over 14 per cent matches increases in exports (16 per cent) and in manufacturing (12 per cent) more closely than the increase in Gross Domestic Product (eight per cent). Between 1985 and 1990, paper manufacture shot up from 294,000 to 889,000 tonnes, reaching 1.66 million tonnes in 1994, with around 3 million tonnes expected in 1997 (Apichai 1992, *P&PA* 11.1994, *PPI* 10.1993, 7.1995). Pulp production doubled in the decade to 1993, to about 200,000 tonnes yearly, and is expected to reach 815,000 tonnes by 2000 (Rajesh 1995).

Among the companies recently making large new investments in pulp and paper have been the Soon Hua Seng Group, one of Thailand's leading rice traders. In late 1995 the firm's subsidiary Advance Agro brought a new 217,000 tonne-per-year Mitsubishi-Beloit paper machine on stream in east Thailand, with an integrated 175,000 tonne-per-year Sunds Defibrator eucalyptus kraft pulp mill following shortly thereafter. Finance for the US$670 million installation was raised largely in the US and Europe (*FT* 23.11.1994). The firm has also ordered another 200,000 tonne-per-year paper machine from Mitsubishi Heavy Machinery. The Siam Cement Group, a company associated with the Crown Property Bureau and which holds eleven companies producing paper and pulp, has meanwhile recently boosted its production capacity to around one million tonnes per year of pulp, paper and converted products, with three new machines slated to come on line by 1997. Hiang Seng and Panjapol have also added new capacity (*Nation* 14.4.1994; *BP* 13.7.1994; *PPI* 8.1995, 11.1995; Rajesh 1995; *Papermaker* 12.1994). Nationwide, overinvestment pushed containerboard capacity to 50 per cent over demand in 1994, and net paper self-sufficiency is expected to climb to around 117 per cent in 1997, when the country will be producing far more short-fibre (eucalyptus) pulp than its papermakers require (Adul 1993; *P&PA* 11.1994; *PPI* 10.1993, 7.1995; Rajesh 1995). Fresh long-fibre (conifer) pulp, however, cannot be produced in large quantities in the country and will continue to be imported.

As in Indonesia, foreign machinery suppliers are important beneficiaries of the boom, with Nordic, Japanese and North American suppliers again dominating the market. Foreign investment is also significant. Phoenix Pulp and Paper, currently Thailand's largest producer of pulp, with 200,000 tonnes per year, is a venture of Ballarpur, India's largest pulp and paper manufacturer — which holds over 13 per cent of the shares — and the European Overseas Investment Development Corporation, which is now the dominant shareholder. In good years, Phoenix exports over a third of its pulp to such countries as India, South Korea, the Philippines, Japan and the US. Asia Tech, a firm partly owned by the New Zealand-based multinational Brierly Investments (see chapter 8), is meanwhile laying the groundwork for pulp and paper mills by promoting contract farming of *Acacia mangium* pulpwood trees over 32,000 hectares in four provinces of Thailand's Northeast. Taking advantage of local government connections, Asia Tech has also illegally seized over 500 hectares of public seasonally-flooded forest on the banks of the Songkhram River to plant eucalyptus. Japanese and Taiwanese firms, too, have formed local consortia to produce wood chips and pulp. Itochu has joined hands with Sahapattanapibul Group to export eucalyptus chips, and New Oji has ties with Advance Agro, which will export jointly-branded products to the Japanese market. The US's Kimberly-Clark, in addition, has invested in Thai tissue capacity and Korea's Shin Ho in a 100,000 tonne-per-year newsprint mill using recovered materials. Banks such as the UK's Barclays meanwhile help advise firms such as Soon Hua Seng on debt finance and share offerings (*PPI* 11.1995, Lang 1995).

Anticipated future growth in both exports and domestic production, together with the need to keep machines running even in times of surplus, has compelled industry to push for more and more of the country's land area to be planted to pulp trees. *Eucalyptus camaldulensis*, known since the early 1970s to be easily adaptable to Thailand's wide range of habitats, and increasingly attractive to paper manufacturers, has been the tree of choice since the appearance of cheap seed on the market in 1978. By 1992 around 80,000 hectares of eucalyptus had been planted by a wide variety of firms and private owners, mainly in the Northeast and Central regions, and industry has continually lobbied the government for further land, recently claiming it needed another 128,000 hectares under fast-growing trees by 2002. With its eye on increasing exports, the industry association has also requested soft loans with grace periods of six years, reductions in corporate income tax, waivers on sales tax, suspension of import duty on pulp- and paper-making chemicals and machinery, more government-underwritten training and infrastructure, and state collaboration in plantation ventures (Apichai 1992, *BP* 1.2.1993, *P&PA* 11.1994, Rajesh 1995).

Official collaboration

Until confronted by massive popular resistance, government officials, by and large, have been only too willing to cooperate. In the late 1980s, government agencies were calling for 4.3 million hectares of commercial tree plantations in National Reserve Forests (or over eight per cent of the country's land area), with communities and the government planting an additional 1.85 million hectares. This eagerness caused some head-scratching among corporate consultants, who, try as they might, could not locate a prospective market for the products of more than ten per cent of such a huge plantation area (Sargent 1990a, *Prachachart Thurakit* 4–6.1.1989).

Such official enthusiasm for commercial eucalyptus plantations, however, is not as mysterious as it may look. The Royal Forest Department (RFD) has been overwhelmingly oriented toward commercial exploitation during its entire 100-year history. When logging was banned in 1989, it was only natural that it should turn its attention to promoting commercial plantations. This bias has been reinforced by the associations many RFD bureaucrats enjoy with eucalyptus nursery owners or brokers and the equally close collaboration the state Forestry Industry Organization (FIO) has forged with the private pulp and paper sector (Apichai 1992).

Promoting commercial plantations has also allowed the RFD to portray itself as discharging its responsibility to 'reforest the country' in the wake of the catastrophic logging of the past 30 to 40 years. This is particularly important in that most RFD officials have consistently held that conservation and restoration of forest cover are beyond the abilities of local villagers, and can only be made possible through state or commercial projects. The RFD has in addition been eager to regain control over large areas of National Reserve Forest land, which, following decades of commercial logging and official promotion of upland export crops such as kenaf, maize, sugar cane and cassava, fell into the *de facto* possession of peasant 'encroachers' (Usher 1990a). Just as the RFD once granted timber companies cheap logging concessions, it subsequently began to grant even cheaper concessions to eucalyptus firms in the hope that they would help drive out the settlers that the logging concessions had helped attract.

The parastatal Forest Industry Organization, too, has been seeking a new role for itself since logging was banned, and as a result has become involved in pulpwood schemes such as commercial plantations on state land as well as in pulp mills. One such mill was mooted for Sri Sa Ket until local opposition forced the FIO to shift the projected site to Ubon Ratchathanee; another is planned for Sakon Nakhon.

Both foreign and domestic eucalyptus-growing firms, moreover, have been linked with political parties responsible for powerful ministries. Soon Hua Seng, for example, whose board chairman is Narong Mahanond, a former Director General of the Police Department, has helped finance the Democrat Party, which for years held the portfolio of the Ministry of Agriculture and Agricultural Cooperatives, which is responsible for the RFD. Asia Tech's land grabs in northeast Thailand, similarly, have reportedly been facilitated by politicians in the New Aspiration, Chart Pattana, Nam Thai and Chart Thai parties (*Krungthep Thurakit* 15.2.1996).

As in Indonesia, in addition, planting trees may not be the only thing tycoons have had in mind when they demand access to more land for pulpwood plantations. While natural forests are not a leading source of pulp raw material in Thailand, businesses have often hired or encouraged villagers to clear forest so that it can be categorized as 'degraded land' legally eligible for planting with eucalyptus. The timber has then been sold illegally for lumber. On occasion, a forest which is not 'degraded' at all may be earmarked for plantations in order that it may be logged (*Nation* 20.1.1996). By putting on a show of interest in pulp and paper, moreover, some business figures have attempted to amass land with potential for profitable uses such as golf courses or tourist resorts. RFD officials and politicians who stand to share the benefits have often been only too happy to play along.

Influenced by Western models, Thailand's official institutions are structured in a way which has helped to subsidize the plantation industry. The Royal Forest Department, for example, has legal jurisdiction over approximately 40 per cent of the country's surface area in the form of National Reserve Forests (NRFs), many of which are forested in name only. Until recently, transferrable land documents could not legally be granted in NRFs, even in the 5.6 million hectares which are estimated to be currently under cultivation, though many NRFs were gazetted in areas already occupied by villagers. In the 1980s and early 1990s, this arrangement enabled the government to label the more than 10 million people who live in the NRFs — some of whom have been present since before the NRFs were gazetted — as 'illegal encroachers'. Accordingly, the land they were occupying could be given away to plantation or other businesses at the derisory rate of US$2.50 per hectare per year, little more than five per cent of the (already low) typical going market rate (Apichai 1992). In theory, the Director-General of the RFD can still authorize the rental of up to 16 hectares of NRF per company, the Minister of Agriculture up to 320, and the Cabinet, still larger areas. In September 1994, the Minister of Agriculture was even given authority to allow state agencies to use parts of protected areas without cabinet approval 'if their projects involve national and economic security'.

Other apparatuses grant special privileges to pulpwood plantations which no other agricultural crop enjoys. In the late 1980s, the RFD set up a separate office devoted specifically to promoting commercial tree farms, complete with a public relations budget of over US$24 million, and in 1994 asked for approximately US$130 million from the 1995 fiscal budget to subsidize private sector plantations and other 'reafforestation' projects, including one in which farmers in National Reserve Forests are given soft loans to plant fast-growing trees. The Board of Investment has meanwhile granted some firms tax holidays and exemptions from duty on imported machinery and raw materials (Apichai 1992, Usher 1990b, *Nation* 9.4.1990, Tunya 1990, *BP* 31.8.94).

Foreign subsidies

Providing additional subsidies are foreign governments. Particularly prominent, unsurprisingly, are those of Japan, Canada, and Finland, three of the countries which are likely to benefit the most either from sales of machinery and consultancies or (in the case of Japan) also from imports of Thai chips and pulp.

As long ago as 1981, the Japan International Cooperation Agency set up a trial eucalyptus plantation in northeast Thailand to foster research and training in the field, and in the early 1990s funded a nursery project growing 20 million tree seedlings yearly. Japanese taxpayers' money has also been channeled through the Overseas Economic Cooperation Fund to support farmers' participation in the Thai-Japan Reforestation and Wood Industry Co., which was designed to supply raw materials to a consortium of Japanese papermakers (Nectoux and Kuroda 1989, Tunya 1990, Masaki 1995). CIDA, the Canadian 'aid' body, has helped finance the Canadian consultant firm H. A. Simons' work with Soon Hua Seng as well as plantation research by the Thai Development Research Institute. Britain's Commonwealth Development Corporation, which draws money from the British 'aid' programme, has provided loans and debt finance to Soon Hua Seng. And the Swedish Board for Investment and Technical Support has hired Swedish consultants Swedforest to help FIO transform itself from a logging to a plantation-managing agency (Usher 1994, Rajesh 1995).

The Finnish government, meanwhile, has pumped millions of dollars of its taxpayers' money not only into export credits for Finnish forestry and pulp and paper equipment, but also into a political lobbying exercise known as the Thai Forestry Sector Master Plan, which was conducted by the Jaakko Pöyry consulting firm between 1990 and 1994. Largely aimed, in the words of its Finnish team leader, at bringing Thailand's 'institutional and social frame into shape', the Master Plan included an ambitious attempt to rewrite the country's entire forestry policy to allow Western techniques of industrial forestry to be applied more fully (see below) (Läitaläinen 1992). While it did not achieve all of its aims, the Master Plan, like similar exercises in Indonesia, did help weave new personal networks among local and foreign plantation proponents in the private, academic and bureaucratic sectors, as well as add to the technical base for national plantation development. The plan also provided a free education in Thai politics for Finnish industry figures which is already paying rich dividends. Finnish machinery exports to Thailand — much of it in the forestry industry sector — rose from US$19 million in 1990 to $113 million three years later, and a special Thai–Finn pact for economic cooperation was signed in October 1993 (Finland National Board of Customs 1990–3). Finnish taxpayers' money has also been used to subsidize the export of Finnish industrial forestry ideology to Thai academic institutions for three decades, in the form of forestry scholarships for Thai students in Finland and Finnish-Thai research projects on industrial plantation species in Thailand (Luukkanen and Hakulinen 1991).

A rural development project funded by the Australian government meanwhile led to much planting of eucalyptus on communal land in the Northeast between 1976 and 1989 under the heading of 'upland rehabilitation'. The project was criticized heavily by local people, NGOs and academics since its beginnings, and its deleterious effects were still being felt in 1996, where local movements in provinces such as Roi Et have been pushing to have eucalyptus removed entirely from rural areas (see below). Yet in 1988 the project won the Australian government's prestigious bicentennial award for 'Excellence in Overseas Development Assistance', and no comprehensive review of the project's failings has ever been undertaken.

Also significant in the industrial plantation boom has been the strong support of multilateral agencies such as the Asian Development Bank, which has funded plantation research and project development in both Thailand and neighbouring Laos. The claim of the Food and Agriculture Organization (FAO) that Thailand should set a goal of 40 per cent forest cover, meanwhile, helps legitimize the notion propagated by plantation interests and the government that 15 per cent of the country should be covered by tree farms.

Popular resistance and its effects

Arrayed against such pro-plantation forces have been well-publicized popular movements, most notably among the more than ten million people living in cleared areas of National Reserve Forests. Since the mid-1980s, small farmers opposing the spread of eucalyptus plantations have petitioned district officials, members of parliament, and cabinet members; held rallies; spoken out at national-level seminars; blocked roads; and marched on government offices. Where other means have failed and they are organized well enough, they have ripped out eucalyptus seedlings, chopped down eucalyptus trees, stopped bulldozers, and burned nurseries and equipment.

Such actions have been motivated mainly by experience showing that large-scale eucalyptus plantations displace or impoverish farmers, violating a strong felt right to a stable subsistence. Even where villagers are allowed to stay on or near plantations, *Eucalyptus camaldulensis* plantations allow little intercropping (villagers say the tree is *hen gae tua* ['selfish'] in that it hogs nutrients), are useless for fodder, supply little firewood to the community, can lower water levels in nearby ponds, wells and woodlands, and can cause erosion which deposits sand in neighbouring fields. In addition, they provide few of the varied forest goods that rural dwellers use to eke out a subsistence, while usurping community woodlands that do. In Thailand's Northeast, where eucalyptus has replaced forests which serve as fish spawning grounds during yearly floods, the tree has even been implicated in recent fishery declines which have halved some village incomes.

Plantation jobs, moreover, are neither plentiful nor steady, and the work is often unattractive to farmers. Land speculation associated with plantation expansion has also often undercut farmer security. Using fraud and legal chicanery and their political and bureaucratic connections, speculators hoping to resell land to plantation companies have threatened rural dwellers with violence, murder, or eviction on 'encroachment' charges; attempted to co-opt village elders; tried to get villagers into debt; or cut off their access to their land by buying up surrounding plots.

Migrating away from plantation areas, moreover, is not an attractive option. In Thailand the frontier is nearly gone — forest cover has declined from perhaps 70 per cent of the country's land area earlier in the century to about 15 per cent today, and government bureaucracies are less likely now to acquiesce to illegal forest colonization than they were in the 1970s, when the country's unsettled upland areas served as welcome safety-valves relieving potentially explosive land conflicts elsewhere. Official resettlement programmes, meanwhile, faced with a dearth of suitable unused agricultural land, have often tried to relocate villagers on already-occupied land, touching off bitter land disputes. Moving permanently to the cities, similarly, is seldom seen as an alluring alternative, and casual agricultural labourers' jobs are insecure and ill-paid. Awareness of their limited choices has stoked villagers' moral outrage at the abuses of the business-bureaucracy-politics patronage system which has been used to promote plantations.

In their struggle, villagers have made shrewd use of a wide variety of rhetorical, political and cultural tools. In many areas, villagers have wrapped their actions in a religious mantle by tying yellow Buddhist monks' robes around large trees to 'ordain' and thus protect them from being cleared to make way for plantations. They have also adapted the *pha pa ba* ceremony traditionally used to mobilize assistance for Buddhist temple construction to promote the anti-plantation movement instead. Well aware of the need to seize the environmental high ground, village plantation opponents have also planted fruit, rubber and native forest trees to pre-empt or replace eucalyptus, or have explained to sympathetic journalists and officials the methods by which they have preserved diverse local forest patches for generations as sustainable sources of water, mushrooms, vegetables, small game, honey, resins, fruit, firewood, fodder, herbal medicines, frogs and edible insects and insect eggs, or as sacred ancestral woods and funeral grounds. Where such traditions have fallen into disuse, villagers have often created community forest conservation organizations to counter the spread of eucalyptus. Examples of naturally forested areas which villagers have 'brought back' after eliminating eucalyptus have also started to appear.

Throughout, villagers have also been careful to stress positive demands for individual land rights, community rights to local forests, and the right to veto commercial plantation schemes slated for their localities. In their struggle, villagers opposing pulpwood plantations have made cautious use of the support of Thai environmentalists and other activists among the intelligentsia, who have called for recognition of local land and forest rights, sharp separation of plantation and forest conservation policies, and cancellation of privileges for industrial plantations. Environmentalists have been particularly helpful in driving home the message that huge increases in commercial plantation acreage can only increase the pressure on natural forests, through displacement and disruption of locally-developed forest conservation practices (Lohmann 1991).

The struggles of the 1990s

In the early 1990s, plantation proponents and opponents alike had to struggle for purchase on a slippery, constantly changing political field. In May 1990, then Prime Minister Chatichai Choonhavan prohibited all commercial 'reforestation' in National Reserve Forests following the arrest of 156 Soon Hua Seng employees for logging a plantation site in eastern Thailand. (In accord with normal practice, Soon Hua Seng had been allowed unofficially to have access to the forest in order to 'degrade' it before negotiations for the concession were concluded, but political intrigue against the company's president led to exposure.) In February 1991, however, a military *coup d'état* paved the way for a massive official programme, which, with a budget of hundreds of millions of dollars, was aimed at evicting five million residents of National Reserve Forests in order to free up approximately 1.37 million hectares for private-sector tree plantations. Within a year, perhaps 40,000 families were forcibly displaced in the Northeast, with repression especially severe in areas targeted by the military as hotbeds of resistance (Sanitsuda 1992, *Phuu Jatkaan Raai Sapdaa* 16–23.9.1991).

When a broad-based popular movement overthrew the military junta in May 1992, resistance to pulpwood plantations emerged again in force. Following major demonstrations by Northeastern villagers, including the blockade of the region's principal highway, and prolonged negotiation with farmer leaders, the government scrapped the military's eviction programme, suspended 'reforestation' with eucalyptus, and imposed a ceiling of eight hectares on any type of commercial tree plantation.

The protests of the last decade have made Thailand a watchword in the industry for conflict over raw materials. This has discouraged both foreign and domestic investment, particularly in giant export schemes such as those to be found in Brazil and Indonesia. Shell, for example, was forced to drop plans to plant 12,500 hectares of eucalyptus in east Thailand in 1990 after violent conflicts and scandals forced delays in governmental approval for the project. Soon Hua Seng, Birla, and other firms have also been compelled to abandon ambitious plantation or pulp schemes out of fears of local opposition or subsequent rejection by the government. Phoenix's mill, meanwhile, lost 141 production days during 1992–4 due to controversies over pollution releases which have damaged local fishing livelihoods, while investors such as Panjapol and Siam Pulp and Paper have run into problems with licensing authorities. Such difficulties have increased investors' reluctance to move out of the Bangkok area, with its good infrastructure, consumers, and easy access to waste paper for raw material (*P&PA* 11.1994).

In September 1993, under pressure from the pulp and paper industry and its allies, Thailand's economic ministers decided once again to open National Reserve Forests to commercial reforestation. In practice, however, grassroots resistance and public opinion have ensured that the conversion of state land to pulpwood plantations remains difficult. In 1994, for example, local opposition to a eucalyptus-planting RFD development programme in Roi Et become so strong that district officials had no choice but to express support for villagers who chopped down over 300 hectares of the Australian tree in order to replace it with community-conserved forests of native species (*BP* 27.3.1995). By 1995, village networks in the province were attempting to eliminate eucalyptus from their areas altogether, forcing the RFD to suspend its eucalyptus operations over a wide area. Responsibility for existing plantations, meanwhile, was passed to other authorities, whom villagers have pressured to cut the eucalyptus and distribute the profits locally. Throughout their campaigns, Northeastern villagers and their NGO allies have researched and publicized multi-purpose native alternatives to eucalyptus which are responsive to the diversity of food, construction, medicinal and ecological needs of different localities; launched supplementary plantings of native trees on degraded sites; and posted new areas as community forest.

Increasingly, government agencies are conceding both the necessity of granting land rights to occupants of NRFs and the problems of large-scale eucalyptus cultivation. In 1994, a new 'land reform' programme called Sor Por Kor was instituted to distribute over 600,000 hectares of NRF land per year to farmers before it was derailed by high-level abuses. A scheme in which private corporations and state enterprises are being invited by the Ministry of Agriculture and Agricultural Cooperatives to help cover at least 540,000 hectares of land with trees between 1994 and 1996 to mark the 50th anniversary of the King's coronation has meanwhile specified that the plantings must be for conservation purposes, must use native trees, and must not evict occupants of the land affected — although whether these goals are achievable is open to question (*BP* 20.1.1995).

From planting leased land to contract farming

Facing clashes with landless farmers or governmental vacillation or opposition in its efforts to take over NRF land outright, the industry and its official allies have turned increasingly to contract farming as a second-choice strategy. Here the industry, instead of dispossessing peasants directly, contracts with them to grow pulpwood trees on their own land, often distributing free seedlings and promising to purchase their harvest years hence at a guaranteed price.

Although this approach can be unwieldy, requiring pulp or chip firms to deal with thousands of smallholders scattered across a large area, it is unlikely to provoke the type of organized resistance to plantations which has followed on from attempts at outright eviction, and may even encourage farmers to clear new areas to plant trees. Thus by early 1996, Asia Tech had convinced farmers in Thailand's Northeast to plant acacia on 16,000 hectares of their own land, half what the firm requires to feed its projected 150,000 tonne-per-year pulp mill in Nong Khai. Soon Hua Seng's Advance Agro, meanwhile, is promoting contract farming among over 4,000 farmers near its mills in eastern Thailand. And Phoenix contracts with over 10,000 farmers within a radius of 150 kilometres for bamboo and eucalyptus supplies; in the area surrounding the company's plant, farmers hard-pressed to find lucrative crops for their sandy soil have been successfully wooed with offers of cash stipends of US$125 per year per hectare for looking after eucalyptus plantings on their land (*PPI* 11.1995, Anuchit 1995b).

While contract tree farming can be more expensive for pulp firms than plantations on leased land, it too is capable of attracting substantial state subsidies. The Sor Por Kor land reform programme, for instance, was associated with a programme providing soft loans for farmers planting fast-growing trees on their new land (*Nation* 15.2.1994). Asia Tech, meanwhile, can rely on the assistance of the government's Agricultural Extension Department in persuading farmers to switch to tree crops. In addition, contract farming is capable of displacing considerable risk onto farmers themselves. Dependent on plantation or pulp firms for seedlings, materials and cash, small farmers may well find out too late that eucalyptus is economically unviable for them and be pushed into foreclosure (Usher 1990b, *Phuu Jatkaan* 3.5.1995, PRED 1996). Plantation or pulp firms may then simply be able to buy up their land at bargain prices.

In the long term, indeed, the types of social and economic control inherent in the contract farming system may well engender new forms of popular opposition to industrial pulpwood cultivation. Over 90 per cent of surveyed farmers participating in the contract farming package

Phoenix has recently pioneered on 325 hectares near its mill in Khon Kaen, for example, already want the project to be suspended (Anuchit 1995a). The package, dubbed 'Project Green' by the company, was conceived partly as a way for Phoenix, plagued by accusations of water pollution, to reduce the amount of (supposedly fully treated) waste water released into the Pong River by using it instead to irrigate the eucalyptus crops of nearby contract farmers. Participating farmers, who were approached as individuals rather than collectives in order to undercut village unity, were entitled to the benefits of the programme — including a guaranteed price of US$32 per tonne for the five-year-old trees — only if they accepted the effluent as irrigation water. Although according to the terms of the contract (of which the only copies are held by the company), participating farmers were to have complete control over the water releases themselves, in reality a team of Phoenix employees circulate among farmers' fields, ensuring that pipes remain open around the clock. As a result, the effluent has not only contaminated and salinated the soil on which eucalyptus was growing, but has also seeped into lower-lying rice fields of non-participating farmers, killing seedlings, full-grown plants, and trees. While the company has been forced to hand over compensation of US$100,000 to villagers, calling it fully to account for the permanent damage it may have done to smallholders' land is difficult (*BP* 12.7.1995, 30.7.1995, 9.8.1995; Anuchit 1995a, 1995b; Wannasri 1994).

Given the problems connected with both contract farming and establishing plantations on state land, some firms have found it advantageous to grow pulpwood trees on land they already own, or to buy up smallholdings now used for other crops. Siam Cement Group, for example, which is the parent company of Siam Pulp and Paper and Siam Cellulose, has recently acquired 1,600 hectares in the north of the country, and Soon Hua Seng has for many years been buying parcels of land from indebted cassava or sugar cane farmers in the east to supplement its contract farming and other plantation efforts. Phoenix's raw materials chief, meanwhile, confessed that he would rather the company buy up smallholders' land adjoining its mill than continue with the Project Green contract farming scheme (Suppachai 1996).

Moving abroad

Another strategy for Thai pulp and paper investors facing domestic resistance is to move abroad. Phoenix, for example, has been working with the Asian Development Bank (ADB), Jaakko Pöyry, the Finnish government, and the European Overseas Investment Development Corporation and its Lao and Swedish subcontractors to develop plans to tap over 10,000 hectares of plantations in Laos. Some 70 per cent of an ADB loan for Lao plantations is earmarked for the private sector — which the Lao state is ill-equipped to monitor — with only 30 per cent to small farmers. Finnish taxpayers are subsidizing the associated planning and political manipulation through a US$5.8 million grant for technical assistance. The grant supports the World Bank's so-called Forest Management and Conservation Project in Laos, which will lead to tens of millions of dollars being poured into a sector where government officials' salaries are around US$30 per month (Malee 1994, Geary 1996). Siam Pulp and Paper, meanwhile, has approached the Laotian government about a possible $250 million project involving a 150,000 tonne-per-year pulp mill and 32,000 hectares of plantations.

Asia Tech, too, has ambitions to invest in plantations in Laos and Burma. A 16,000-hectare *Pinus radiata* plantation planned for Laos's Bolovens Plateau, according to Asia Tech's own estimates, will displace nearly 5,000 people, for whom no new land has yet been found. A bit over US$1 per person will be allocated during the project's first year toward finding new livelihoods for the evictees. Huge quantities of pesticides, herbicides, fungicides, and chemical fertilizers will be used on the project area, including glyphosate, Pulse, Simazine and Gardoprim. A mere US$3 per hectare will be paid in annual rent to the state during the first ten years of the project (Bannan 1995). Asia Tech is also planning to cooperate with the Lao government and a New Zealand-based Jaakko Pöyry affiliate in another, far larger acacia plantation project in Khammouane and Kham Xai provinces. A pulp mill to be powered by one of the destructive new dams being built on Mekong tributaries is also contemplated (*Krungthep Thurakit* 15.2.1996, Geary 1996). The Lao military logging company Pattanakit Poudoy, in addition, hopes to build a pulp and paper mill in Vientiane province which would be supplied by a 20,000 hectare plantation and a one million hectare bamboo forest concession. Bidders for the project include the European Overseas Investment Development Corporation, and Phoenix is expected to help with mill operations (Geary 1996).

Meanwhile, like Indonesia's Sinar Mas and Raja Garuda Mas, Soon Hua Seng has moved into China, investing hundreds of millions of dollars in joint ventures to plant over 220,000 hectares of eucalyptus in Canton and set up chip, pulp and paper mills. Led by the Ministry of Agriculture and Agricultural Cooperatives, the Thai government has also signed a cooperation pact on industrial forestry science and technology with China which is aimed at, among other things, supporting that country's policy of planting 6.5 million hectares of fast-growing trees such as eucalyptus and poplar by the year 2000 with investment from Japan, New Zealand, Australia, Indonesia and Thailand (*BP* 23.6.1993, Rajesh 1995).

Siam Pulp and Paper, in addition, is taking advantage of Indonesia's 'huge wood resources' by buying into a US$1 billion, 350,000 tonne-per-year pulp joint venture in east Kalimantan with the Astra International Group *(FT* 4.1.1996*)*. Siam is also investing in a kraft paper mill in the Philippines with Phinma Group *(Nation* 1.3.1996*)*.

The Thai Forestry Sector Master Plan: a case study of international pulp and paper politics

Jaakko Pöyry's Thai Forestry Sector Master Plan (TFSMP) offers an interesting illustration of the attempt by the international pulp and paper industry and its allies to build up support networks, rewrite policy, and manage resistance in an intercultural context.

Much of the original impetus for the TFSMP, like that for other national Forestry Master Plans, came from the Tropical Forests Action Programme (TFAP), a gigantic scheme which had originated in an early 1980s conversation in a Washington bar between the World Bank forester John Spears and a colleague, who were brainstorming ways of getting more international development funding for professional forestry consultants in the new atmosphere of concern in world capitals for tropical forests. As developed by the World Bank, FAO, the United Nations Development Programme (UNDP) and the World Resources Institute, the plan encourages each tropical forest country to join donor agencies such as CIDA, FINNIDA or the World Bank in setting up missions to review the state of its forests. Management plans are then formulated — usually highly commercially-oriented — whose components can attract funding from international or bilateral agencies (Colchester and Lohmann 1990).

A second set of interests behind the TFSMP was Pöyry's consultancy business (see chapter 4). Having helped to establish Pöyry as a key player in several Asian countries, including the Philippines, Indonesia and Nepal, Jouko Virta, President of the firm's Consulting Division, was hard at work in the mid-1980s trying to make inroads into the highly personalized Thai government system. Virta's way was smoothed by a fortuitous meeting with a Swiss named Nat Inthakan, who had been living in Thailand for several decades and had Thai nationality and an intimate knowledge of the local timber industry. Nat arranged introductions for Virta to Snoh Unakul, a businessman and Secretary General of the National Economic and Social Development Board; General Harn Leenanonda, then Minister of Agriculture and Agricultural Cooperatives, which had jurisdiction over the Royal Forest Department (RFD); and Phairote Suwannakorn, then Deputy Director of the RFD. Virta then wrote up terms of reference for a Master Plan for Thai forestry development which Nat, now acting as representative of Pöyry in Bangkok, used to brief then Prime Minister Prem Tinsulanonda. On a 1988 visit to Finland, Prem signed an agreement whereby FINNIDA would fund a Thai master plan along TFAP lines. Since Thailand's per capita Gross National Product was too high for the country to qualify technically for Finnish bilateral aid, FINNIDA's funds were channeled through UNDP. UNDP duly selected Pöyry as plan consultant. Rauno Läitaläinen, who had been in charge of Pöyry's master-plan team in Nepal, arrived in Bangkok in July 1990 with a tax-free annual salary of US$240,000 (Usher 1991).

Despite Prem's imprimateur, Läitaläinen faced immediate problems in creating a plan which could satisfy the various bureaucracies, state enterprises, businesspeople and speculators with interests in the forests — to say nothing of farmers' groups, non-governmental organizations and environmentalists. Having had prior experience with FAO and UNDP, and aware of the controversy over TFAP, the Forest Department's planning division at first refused to work with Pöyry, and as late as August 1991, an FAO official found that 75 per cent of Forest Department staff remained opposed to the plan (Inglis 1991). Some 205 NGOs involved in rural development, meanwhile, noting Pöyry's unconcealed interests in promoting commercial plantations, objected to the plan on the grounds that it would strengthen state and industrial control over forests at the expense of local communities and their commons. Under pressure, Läitaläinen agreed to sign a statement stipulating that the plan's Terms of Reference be rewritten after consultations with NGOs working with village communities.

Läitaläinen and his team then began to devote time to learning the political ropes, lobbying for a coordinated approach to industrial forestry, making elite alliances, distributing consultancies, starting up publications, mollifying malcontents, and lining up potential supporters for a plan many of whose details would be left for the future. Perhaps sensing that the more participants he brought in on the side of the plan, the easier it would be to accuse others of 'marginality' and 'obstructionism' and to pass responsibility to the authorities, Läitaläinen went out of his way to make the TFSMP seem capable of answering the needs of all actors. When speaking with the government's Forest Department, he suggested that TFSMP could help increase the country's forested area and wood industries. When speaking with business, Läitaläinen stressed the need for the government to subsidize private investment in plantations through provision of land and other necessities. Trying to integrate NGOs into the planning process, he praised grassroots efforts to conserve forests; acknowledged the need for land reform, popular 'participation', and grassroots benefits; and claimed to be in a unique position to intervene with the state on behalf of ordinary people. Photographs of villagers sitting in conclave or planting trees under the tutelage of officials began to feature in TFSMP documents. The TFSMP was presented as an infinitely self-correcting 'rolling process' capable of accommodating any objections from any actor. Criticisms were dismissed as 'premature' and critics invited to participate in succeeding stages.

These efforts had some effect. The TFSMP's ability to hire consultants eventually attracted many Thai forestry faculty. Although most Thai NGOs stuck to their demand that the Terms of Reference of the plan would have to be revised before they would consider any invitation to participate in the planning process, two NGOs agreed to serve on the steering committee in the spring of 1991. One of these soon afterwards received an unusual US$20,000 grant from FINNIDA.

Pressure nonetheless built from the majority of NGOs, who pointed out that Pöyry, in providing supposedly 'neutral' information about economic demand and forest resources and uses, promoting 'correct' management techniques and environmentally friendly technologies, trying to integrate land managment into global wood-fibre supply systems, and proposing repeal of the popular 1989 logging ban, was in fact already engaged in political subversion against land reform and many village ways of life. The planners' profession of support for customary land rights and local control meant little, NGOs noted, given that, under the plan, villagers' own systems of knowledge and organization were to be subordinated to technocrats' schemes and 'measures undertaken to . . . accelerate out-migration from the forest lands'. Jaakko Pöyry consultants' 'bottom-up' planning, they added, was bottom-up in name only, since in fact it consisted merely in officals' 'outlining' their management plans in the presence of villagers.

Partly out of reaction to such pressures, and partly out of a typically corporate frustration with Thailand's existing 'institutional and social frame', the master plan team moved further and further away from presenting itself as a mere 'technical' appendage supplying facts to a unified body of forward-looking policymakers. Instead, it was forced to begin advertising itself as a political facilitator of a compromise 'national vision' of Thai forests, a reservoir of expertise on democracy and 'participation', and a redrafter of policy. Predictably, this stance roused even sharper sarcasms. As one NGO leader noted in a 1993 letter to the *Bangkok Post*, '"National values" as perceived by the master plan team bear little resemblance to the values local people place on collectively managing community forests and commons within cultural, social and economic contexts of local communities throughout the Kingdom'. In the end, the company was forced to cut its losses with

NGOs. Allegedly on the suggestion of Heikki Rissanen, forestry adviser to FINNIDA, Läitaläinen broke his promise to sign the recommendations coming out of the February 1991 meeting (Wallgren 1994). Jouko Virta, although aware that the bulk of Thai NGOs involved in rural work opposed the TFSMP, went on record claiming that only two or three marginal and 'extremist' individuals — 'I think they are anarchists' — were critical of Pöyry's planning exercises. The claim began to be heard that it was 'too late' to influence the plan and that any problems with it were due to NGOs' refusal to participate.

Yet Thai officialdom proved hardly more willing than NGOs to indulge Pöyry's pretence of being able to redraft Thailand's entire forest policy and reform its practice from top to bottom. The cabinet never approved the completed TFSMP; nor did any state bureaucracies rally round its banner. Predictably, the plan wound up, in words which anthropologist James Ferguson has used to describe development projects in Lesotho, like a 'bread crumb thrown into an ant's nest' (Ferguson 1994). Instead of providing a blueprint for a brand-new, comprehensive and coherent forest management regime, the plan remained a relatively small component in a larger machine, treated at most as a 'shopping list' from which various actors could choose isolated items which could benefit their own circles.

Pöyry's ability to pursue the master plan at all, and thus to carve out an at least slightly more spacious niche for pulp and paper interests, was due partly to the fact that it could successfully conceal from the Finnish public the scale of resistance its schemes were experiencing in Thailand. In this it was helped not only by the physical distance between the two countries, but also by the fact that the Finnish public shared many of the Pöyry consultants' assumptions. To many Thai observers it was merely common sense that Pöyry, in laying out the master plan, was seeking commercial benefit and that it was unaccountable to the people whose livelihoods it was threatening. In Finland, where the belief in the 'neutrality' of corporate consultants and their 'objective expertise' remained strong, such claims, if they could be heard at all, often sounded like paranoia. Similarly, to many Thais, the idea that Finnish foresters could provide a neutral forum in which the goals of (say) transnational corporations, Thai government ministries, local politicians, and Northeastern villagers could be reconciled under centralized authority seemed fanciful. In Finland, however, it was given solemn credence. Finally, while in Thailand it was common knowledge that millions of rural residents depended for their livelihoods on the type of commons regimes commercial eucalyptus schemes were disrupting, such regimes seemed merely quaint or economically marginal to many Finns.

Chapter 13

Conclusion: Looking to the Future

The priority of politics

As this book has shown, a great deal of bad or inappropriate science and bad or inappropriate development economics is used to promote large-scale industrial tree monocultures. But these plantations are not established simply out of an intellectual mistake. They do not proliferate merely because someone, somewhere, has made a scientific error or perpetrated a fraud in biology, accounting, or risk analysis.

Plantations are established, rather, because they serve certain interests, make sense within certain discourses, and provide a whole range of organizations with opportunities to reshape their surroundings, and themselves, in a way which prolongs their survival. Plantations result in real gains for real people — many of whom may be found in the institutions described in chapter 5. These gains seldom 'trickle down' to the people who live where the plantations are established, and are accompanied by long-term degradation of the land and livelihoods of large numbers of communities. But there is little use in denying that they exist.

If the institutions of chapter 5 did not exist, few people would pursue, and still fewer would accept, the one-sided science and economics used to promote massive industrial plantations. If academic forestry were not so closely confined, historically and institutionally, within the conceptual universe which has grown up around the wood industry, fewer foresters would be likely to neglect the issues this book has highlighted, or to make excuses for the industry's depredations, or to give short shrift to existing or potential alternatives, and more foresters would be likely to recognize the possibility of finding common cause with more diverse groups. Similarly, if no institution were assigned the job of finding taxpayer subsidies for big, destructive pulp mills, the biased economics which justifies the operation would perhaps not need to be promoted so shrilly as 'neutral' social science.

But as long as such institutions enjoy disproportionate power, bad forestry science and bad development economics will continue to be produced in great quantities, no matter how 'logically' they may be refuted in scientific papers and actual plantations. This flow will hardly be stemmed merely by herding those responsible into a seminar room and pointing out a few scientific errors or omissions. The interests and institutions which feed it must also be challenged — and not only in meeting rooms but also in fields and forests — by political alliances including those whose resources industrial plantations have stolen.

By the same token, any attempt to promote 'solutions' to the patterns of destruction outlined in this book is unlikely to bear fruit if it is undertaken exclusively within the conceptual and political universe of the institutions examined in chapter 5. Many practical 'alternatives', indeed, are already well-known, but because they are not particularly friendly to the institutional cultures involved, they have been neglected.

It is well-established, for example, that a reasonable level of paper use does not intrinsically require huge machines, large technocracies, extensive road networks, intercontinental marketing mechanisms, the mining of vast amounts of raw material in single locations, or the immensely water-, energy-, chemical- and capital-intensive installations which characterize the mainstream industry today. China, for instance, still supplies its immense paper needs largely through small local mills which use surplus local agricultural wastes such as straw, support community economies, require no advanced infrastructure to support them, and, like village bakeries, can safely shut up shop temporarily when no one is buying without the proprietors needing to worry about paying off their machinery investments. While little effort has been dedicated to effluent treatment, with serious pollution problems a result, there are no overwhelming technical or economic obstacles to running such mills cleanly (Wong 1992).

Paper manufacturing expert A. W. Western (1979), moreover, has argued that in India and other Southern countries, 'detailed comparisons between the large mill and the equivalent capacity in small mills overwhelmingly favour the smaller unit in economic terms'. According to researcher Maureen Smith (1995), there are no serious obstacles even to current US paper consumption being met by a more decentralized network of small- to medium-sized mills using a raw material base of approximately half waste paper and half non-wood crops including straw, hemp, or other regionally-appropriate materials.

Much information is available, too, showing how excessive pulpwood use by the North can be avoided. The Rainforest Action Network in San Francisco, for example, has put out an informative and well-thought-out booklet entitled *Cut Waste, Not Trees: How to Use Less Wood, Cut Pollution and Create Jobs* (available from RAN at 450 Sansome Street, Suite 700, San Francisco, CA 94111, USA); and Friends of the Earth UK has produced an outstanding book entitled *Out of the Woods*, which outlines concrete steps for reducing pulpwood consumption in Britain (available from FoE at 26–28 Underwood Street, London N1, UK).

None of these paths, however, are likely to be taken by industry, or supported by most governments, without strong social and political pressures for more democratic control of the paper economy being brought to bear — which is why the last-mentioned books have been explicitly designed as part of extended campaigns.

Nor are other, newer paths likely to be found merely by researching 'alternatives' and experimenting with technical fixes while leaving for others the political question of who is to implement them and how. Doing so is likely to lead, by default, to those 'alternatives' being constrained, perverted and deployed by these same institutions in such a way that they are of little use to industrial plantations' most important critics. That would lead, in turn, to a failure to challenge effectively the prevailing unsustainable forestry model. Indeed, only when inquiries into 'solutions' are conjoined with concrete efforts to cooperate with or encourage popular movements challenging the excessive power of the institutions of chapter 5 does the full range of alternatives which are of most practical use to currently available political forces even become visible.

To be sure, researchers who work closely with industry and shun popular movements will still have incentives to come up with various technical 'solutions' to the 'problems' posed by plantations and paper manufacture. They may explore, for example, timber and paper certification, recycling, alternative fibre sources, programmes to plant a greater variety of trees in plantations, schemes to move plantations to new locations, biotechnology, compensation programmes, chlorine reduction, and water-conserving closed-cycle mills (Smith forthcoming; Galloway 1994; Dudley, Stolton and Jeanrenaud 1995). And many of these 'solutions' will be accompanied by information and analyses which rural plantation critics and environmentalists find tactically useful. Yet to advocate them by themselves — on the grounds that only industry-friendly proposals will have an effect on industry practice — is to forget the lesson that another way of influencing the progress of the games that industry plays is to refuse to play them.

Seekers of more firmly-rooted, lasting, thoroughgoing, and politically-feasible 'alternatives' to industrial plantations, and the forces which give rise to them, will thus also involve themselves in movements for broader social change. They will not rest content merely with demonstrating, for example, that paper can be made from hemp as well as from wood, since without more thoroughgoing changes in industrial structure, some of the same social problems familiar from wood-based papermaking — expropriation of smallholders, for example — could also persist with hemp. Rather, they will also inquire whether the hemp alternative is useful as a political tool to rural tree-plantation opponents and investigate, together with those opponents, how a hemp-based paper economy might be conjoined with more democratic, decentralized, livelihood-friendly and environmentally-stable patterns of land use and paper consumption.

Some of the most interesting practical examples of popular movement-aligned research into 'alternatives' to large industrial plantations can today be found in Thailand, where villagers and NGOs have joined together to document the sustainability of systems of stewardship of land and forests based on local power and a respect for local ecological and social diversity. Researchers allied with the movement of Karen villagers to keep control of their land in north Thailand, for example, are demonstrating the superiority of long-tried systems of communally-managed swidden agriculture to industrial plantations in maintaining tree cover and biodiversity in highland watersheds (Jawni 1996). Similarly, cooperative 'action research' among villagers and NGOs in the country's Northeast is bringing to public attention the virtues of locally-managed plantings of the native trees on which local villagers have long relied (PRED 1996).

One important conclusion emerging from this research is that, in order to promote sustainable 'alternatives', certain premises of the industrial system must be called into question at the outset: for example, that a single tree species must be found which is suitable for many localities, or that 'demand' must be met no matter where it arises or what it is based on. This conclusion is more likely to flow out of inquiry closely engaged with the realities of grassroots livelihood, knowledge, and social action than from the efforts of scientists linked to the entirely different realities of the corporate world.

Only through close contact with popular movements, too, can investigators and activists interested in 'alternatives' discover which proposals would at present be counterproductive, and which processes of political alliance might be necessary to the achievement of other solutions. Only through contact with people aware of the actual grassroots-level results of programmes to certify plantation wood as 'environmentally sound', for example, will NGOs be able to determine whether these programmes are not in fact merely shoring up business as usual, and thus whether it is strategic to participate in them. For those concerned about the problems of the plantation boom this book has described, the first question should not be 'What are the solutions?' but rather 'Who are we working with to find and press for solutions?'

Hence just as Part One of this book insists on presenting facts about plantations and their problems (chapters 1, 2 and 4) together with a discussion of the actors, interests and strategies which promote them (chapters 2, 3, 5 and 6), so this concluding chapter insists on not divorcing talk about 'solutions' from the context of the political battles which are necessary to achieve them. Instead of trying to distill 'policy recom-mendations' or 'position statements' for state, international, or private institutions which are assumed to be ready to act on them, it puts its emphasis, more realistically, on sketching some of the challenges facing existing political alliances among plantation critics.

Working positions

This is not to deny that this book has 'working positions' on various issues as a result of its analysis of the history of pulpwood plantations in a variety of countries; it does. These include the following:

- Large monoculture industrial tree plantations are socially and environmentally unsustainable. There is no place for them in social systems seeking to nurture soil and people.

- Local people must have the right to veto land uses and manufacturing processes they do not accept.

- Ways must be found of promoting or sustaining decentralized pulp and paper manufacture; adjusting it to local needs and plans; reducing the industry's scale and dependence on vast amounts of single, standardized commodities; and lowering demand, particularly in the North.

- Large industrial tree plantations cannot be fruitfully discussed in isolation from the global economic and social realities of which they form a part. The issues they raise are political, not merely technical; for the plantations to be curbed thus requires political action in a broad sense.

Such 'working positions' can be useful in many ways. At the very least, they are a means of organizing the thoughts of plantation critics in ongoing discussions and struggles. They also express conditions which would have to be fulfilled by any international attempt to 'certify' pulpwood timber as environmentally and socially sound; and insofar as enforcement of these conditions is not feasible, they tend to call into question the feasibility of certification. These 'working positions' are best viewed, however, not as master plans which dictate in advance every action which is to be taken against plantations, but merely as occupying a small clarifying role in a large practical social dynamic.

Alliances among interest groups

The struggle against large-scale industrial tree plantations is fought in many different social arenas at the same time: in rural areas, newspaper pages, government offices, informal seminars, test sites, scientific journals, lobbies, courts, parliamentary corridors and economists' conclaves. In each arena, different things may be at stake and different styles of argument used. Yet it is often important to be able to work in all such arenas. This is for at least two reasons. First, it is undemocratic to ask all plantation opponents to inhabit the same arena — to ask rural villagers, for example, to prepare scientific papers proving the chemical effects of eucalyptus or pines on the soil. Nor is it democratic to ask plantation opponents to stand by silently while lobbyists or lawyers pursue their case 'through proper channels'. All critics should have opportunities to express their opposition in arenas of their choice. Second, to restrict opposition to plantations to only a few social arenas is simply not politically effective. Publicity, lobbying, scientific research and grassroots action, undertaken at the right moments as components of a single struggle, will be more effective than any one of these pursued alone. Movements which have many arenas in which to act (as in Thailand) tend to be stronger than those which have few (as in Uruguay or Indonesia).

There are, however, no rules of strategy which dictate when it will be effective for plantation critics to take action in one arena rather than another. For example, going to court against pulp interests (as has happened in Indonesia), or promoting new forest legislation safeguarding the rights of local people (as has happened in Thailand), can be useful at certain moments when newspapers are paying attention to a court case, or when the law is in flux — even when no one believes that the solutions to plantations' destructiveness will ultimately be merely legal. At other times or in other circumstances it may be more useful to sit down and debate forestry consultants, publicize local abuses, or present 'alternatives'. This is a matter of political judgment and consensus best exercised by those closely acquainted with the relevant local and national contexts.

Effectiveness therefore requires alliances between different groups of plantation opponents with different interests and different skills acquired in struggles in different arenas. Only through careful discussion is it possible to find out what sort of scientific, lobbying, or legal action will mesh most effectively with potential or actual struggles in local arenas.

International solidarity

International alliances are both necessary and problematic. While they aim to strengthen local groups, make possible networks of support and advice, and desanctify centralization, they also use 'global' tools of communication, such as electronic mail and international meetings, which by themselves seldom foster much cross-cultural understanding. Effective mutual action may be delayed until areas of common concern and prior misunderstanding are identified. Northern and Southern groups may differ, for example, even about what a plantation is, and some Northern groups receptive to criticism of the expropriation and environmental destruction associated with plantations in the South may not be as strongly critical of plantations in the North.

International alliances may be of many kinds. Southern groups may share information and strategic thinking with other Southern groups within a region or across the globe. Southern groups may also offer insights and solidarity to Northern movements concerned with the protection of local land and forests, as has happened in the Nordic countries, whose growing forest networks have benefited considerably from the lessons learned from struggles in the South.

Northern groups, finally, can also play an important supporting role in Southern groups' attempts to curb the damage done by plantations. They can, for example, monitor the plantation-promoting activities of the bilateral 'aid' agencies, consultancies, commercial development investment agencies and transnational corporations based in their countries. Northerners can legitimately both question the use of their tax money to subsidize destructive commercial plantations abroad, and to challenge the practices of companies which use their countries as headquarters. Thus Finnish NGOs such as the Finnish Forest Action Network, in close association with Southern NGOs, held pathbreaking public forums in the early 1990s in Helsinki at which NGOs from Thailand, Indonesia, the Philippines, Nepal and other countries criticized Jaakko Pöyry and its official helper FINNIDA for interfering in their countries to promote plantations, and called on the Finnish public to curb the abuses of its government agencies and corporations.

Northern groups can also call Northern investors' attention to the destructiveness and the risks connected with Southern private-sector projects floating stock or bond offerings in the North. Thus British activists raised questions about Barito Pacific's operations during its attempts to raise finance in London for its gigantic TEL mill in Sumatra (see chapter 11). The threat of a boycott of Scott Paper products in Europe and North America, similarly, forced the company to withdraw in 1989 from a scheme for a plantation in West Papua which would have displaced thousands.

There is also a great deal of potential for international links between, on the one hand, community and labour groups in the North organizing around issues of contamination with toxics (particularly dioxins) and paper company union-busting, and, on the other, Southern groups concerned about pulp mill and plantation expansion. Both the Citizens' Clearinghouse on Toxic Wastes (CCTW) in the US and the Women's Environmental Network in the UK have launched campaigns on dioxins whose research and activities could benefit, and benefit from, related movements in the South.

Of special potential, perhaps, are links between Southern and Northern regions which are threatened in similar ways by the pulp and paper industry's expansion outside traditional resource areas. As the industry exhausts the resources of regions such as North America's Pacific Northwest, it is not only establishing plantations in the South, but also logging new frontiers in (for example) West Virginia and eastern Siberia. A united front among concerned people in such regions could help check the advance of the industry in each local area.

Sharing of general information can also be useful, whether it is analysis of the possible consequences of technological or market shifts, or news on alternative land and community forest management systems, or findings about the specific biological dangers associated with pine plantations (Rosoman 1994), or ideas for keeping the plantation industry away by increasing the cost of capital and thus the cost of the pulp mills the industry wants to build (Graham 1994). Media for exchange range from news clippings to videos to electronic conferences managed by the Association for Progressive Communications. These last include the conferences for.paper (in English) for.plantation (in Spanish) and dioxins.info (in English). While such information cannot always be adapted for use in local circumstances, sharing it can often stimulate strategic thinking.

In sum, the powerful alliance of interests described in chapter 5 requires an even stronger opposing alliance which recognizes that people are not helpless pawns in the hands of large corporations, states and international agencies. Well-informed and well-organized citizens at local, national and international levels can find common ground and put a halt to the global advance of the currently-dominant model of tree plantations and paper manufacture. Indeed, they are already starting to do so. The aim of this book — whose authors have relied on the views and comments of a large number of people concerned with plantations and their links to the paper industry — has merely been to lend a modest hand to this alliance by sharing some of this information and analysis. We hope that it may prove a useful tool.

Bibliography

Adams, J. A. (1978) 'Long-term Aspects of Nutrient Loss from Forest Soils and Ecosystems', *New Zealand Journal of Forestry* 23 (1).

Adams, John (1995) *Cost-Benefit Analysis: Part of the Problem, Not the Solution*, Green College Centre for Environmental Policy and Understanding: Oxford.

Adams, Patricia (1991) *Odious Debts*, Earthscan: London.

Adlard, Philip G. (1993) *Historical Background*, Study No. 1, *Shell/WWF Tree Plantation Review*, London: Shell International Petroleum Company and World Wide Fund for Nature.

Adul Udol (1993) 'The Development of the Pulp and Paper Industries in Thailand', in *Appita Journal* 46, 1: 14–18.

African National Congress (ANC) (1994) 'Agricultural Policy: 11. Forestry'.

Albertyn, Chris (1994) Personal communication.

Allen, J. (1992) 'Aspirations - and the reality', *Paper*, 4 February.

Anon (n.d.) 'Profile of the Indorayon Pulp and Rayon Mill', ms., Medan.

Anuchit Karakate (1995a) 'Samruaj Thatsanakhati Chaaw Baan kap Project Green' (survey of villagers' views of Project Green), Thai Volunteer Service: Bangkok.

—— (1995b) 'Phon Krathop Taw Chumchon: Koranii Khrongkaan Chai Prayoot jaak Nam Thing Phuea Kaankaset Phattana Aachiip lae Raaidai Chumchon (Project Green)' (effects on the community of Project Green), ms., Bangkok.

Apichai Puntasen, Somboon Siriprachai, and Chaiyuth Punyasavatsut (1992) 'Political Economy of Eucalyptus: Business, Bureaucracy and the Thai Government', *Journal of Contemporary Asia* 22, 2: 187–206.

Aracruz Celulose (1996) *The Eucalyptus and Sustainable Pulp Production*, Rio de Janeiro: Aracruz Celulose.

Argentina Secretaría de Agricultura, Ganadería y Pesca (SAGP) (1992) *Guía forestal*, Buenos Aires: SAGP.

Asian Development Bank (ADB) (1993) Indonesia: Private Sector Industrial Tree Plantations Program: Report of a Technical Assistance Consulting Team, Manila: ADB.

Associação Brasileira de Exportadores de Celulose (ABECEL) (n.d.) *ABECEL*, Rio de Janeiro: ABECEL.

Associação Gaúcha de Proteção do Ambiente Natural (AGAPAN) (1992) 'Dioxins from pulpmill: South Brazil', in haz.pulpmills electronic conference, 11 February.

Associação Nacional dos Fabricantes de Papel e Celulose (1995) *O Relatoria Estatistico Florestal de 1994*: ANFPC.

Association for Progressive Communications (APC) (1992) 'Brazil pulp mill scandal', in haz.pulpmills electronic conference, 10 June.

Aurell, Ron and Jaakko Pöyry (1988) 'Pulp and Paper: Worldwide Trends in Production, Consumption and Manufacturing', in Schreuder, G. (ed.) *Global Issues and Outlook in Pulp and Paper*, Seattle: University of Washington.

Australia, Parliament of the Commonwealth, Parliamentary Research Service (PRS) (1994) *The Woodchip Licensing Issue*, Current Issues Brief 31, Canberra.

Avenor (1993) *Yearly Report*, Vancouver: Avenor.

Axberg, G. N. and P. H. Ståhl, (1989) 'How Much Wood Does Your Forest Yield?', in *Know-How Wire: Jaakko Pöyry Client Magazine*, January: 11–13.

Ayres, Ed (1993) 'Making Paper without Trees', *Worldwatch*, Sept./Oct.: 5–8.

Bainbridge, W.R. (1990) 'Commercial Afforestation in the Approaches Zones of the Natal Drakensberg as a Potential Threat to Environmental Quality and the Integrity of Natural Communities in the Drakensberg region' in University of Natal Institute of Natural Resources, *The Physical, Social and Economic Impacts of Large-Scale Afforestation in Natal/Kwazulu*, Proceedings of the Forestry Impacts Workshop, Pietermaritzburg: Institute of Natural Resources, May.

Ball, J. B. (1992) 'Forest Plantations and the Wise Management of Tropical Forests', in Adam, K. L. and F. R. Miller (eds.) *Proceedings of the Oxford Conference on Tropical Forests: Wise Management*, Oxford: Oxford Forestry Institute.

Bannan, K. J. (1995) *Pre-feasibility Report on a Softwood Plantation Project on the Bolovens Plateau, Champassak Province, Southern Laos*, prepared for Asia Tech Co., Ltd.: Rotorua, NZ: Forenco Consultants.

Barnett, Adrian and Juniper, Tony (1992) *Deserts of Trees: The Environmental and Social Impacts of Large-Scale Reforestation in Response to Global Climate Change*, London: Friends of the Earth.

Bayliss, Martin (1995) 'India's Paper and Board Industry: Steady but Unspectacular', *P&PA* 21, May: 29–39.

Bazett, Michael D. (1993) *Industrial Wood*, Study No. 3, *Shell/WWF Tree Plantation Review*, London: Shell International Petroleum Company and World Wide Fund for Nature.

Beer, Michael (1995) 'Profile of Sappi: South Africa's Major Pulp and Paper Producer Goes Global', *P&PA* 22, August.

Bermejo, Isabel (1994) 'The Conservation of Dehesa Systems in Extremadura' in E. M. Bignal, D. I. McCracken, and D. J. Curtis, *Nature Conservation and Pastoralism in Europe*, Peterborough: Joint Nature Conservation Committee.

—— (1995) 'Background Paper', ms., Vernejo, Cantabria.

Bethlehem, Lael (1994) 'An industrial strategy for the pulp and paper sector', Cape Town: University of Cape Town Press.

Biermann, Christopher J. (1993) *Essentials of Pulping and Papermaking*, San Diego: Academic Press.

Bingham, John (1995) 'The World Pulp Market: Where Next?', *P&PA* 22, August.

Bissio, Roberto (1991) Personal communication.

Brennan, Wendy and Pappens, Rita Anna (1995) 'Annual Review: Producers Set New Records', *PPI*, July: 19–23.

Brooks, Teresa (1994) Personal communication, December.

Brouwer, Roland (1993) 'Between Policy and Politics: The Forestry Services and the Commons in Portugal', *Forest and Conservation History* 37, October: 160–67.

—— (1995) 'Commons, Pulp and Politics: Portuguese Communities in Defense of their Baldios', ms., Wageningen.

—— (1996) *Planting Politics: The Afforestation of the Commons and State Formation in Portugal*, Wageningen: Netherlands Organization for Scientific Research.

Bruno, Kenny (1992) 'O Relatorio Greenpeace sobre a Maquiagem Verde: O Desfarce Ecologico das Empresas Transnacionais', Rio de Janeiro: Greenpeace.

Buxedas, Martín (1995) 'Un gran sistema maderero: notas sobre la realidad y sus posibilidades', Montevideo: CIEDUR.

Cabaña, Carlos (1993) 'Forestación con eucalyptus en regiones VIII y IX: una válvula hacia la reconversión', *Chile Forestal* 211, November.

Cameron, Owen (1994) Personal communication, March.

Campinhos Junior, Edgar (1992) 'Plantación de eucalipto de alto rendimiento, *Uruguay Forestal* 3, June.

Campos Perrone, Edson and Borges Filho, Orlindo (1994) 'A fauna associada ao plantio de eucaliptos', *Ecologia e Deenvolvimento* 3 (41), July.

Cardellichio, P., Yeo Chang Youn, D. M. Adams, Rin Won Joo and J. T. Chmelik, (1989) *A Preliminary Analysis of Timber and Timber Products Production, Consumption, Trade and Prices in the Pacific Rim Until 2000*, Seattle: Centre for International Trade in Forest Products, College of Forest Resources, University of Washington.

Carey, Alex (1995) *Taking the Risk out of Democracy: Propaganda in the US and Australia*, Sydney: University of New South Wales Press.

Carrere, Ricardo (1989) 'El complejo forestal: situación actual y perspectivas', Montevideo: CIEDUR.

—— (1992) 'Forest development and environment in Uruguay. 17: Forest geopolitics and sustainable development. An overview and the Uruguayan example', Montevideo: CIEDUR.

—— (1993) 'Dossier forestación', *Tierra Amiga* 12, April: 15–28.

—— (1994) 'La invasión de las especies exóticas', *Tierra Amiga* 22, March.

Carrere, Ricardo et al. (1995) *Impactos de la Forestación en Uruguay*. Montevideo: ITEM/REDES.

Castroviejo, Bolibar et al. (1985) 'Política forestal en España (1940-1985): ¿Producción o conservación?', *Quercus* 19.

Cavieres, Aaron and Lara, Antonio (1983) 'La destrucción del bosque nativo para ser reemplazado por plantaciones de pino insigne: evaluación y proposiciones I. Estudio de caso en la provincia de Bío-Bío', Santiago: CODEFF.

Centre for Science and Environment (CSE) (1995) *Should Paper and Pulp Industry Get State Forest Lands for Captive Plantation?*, New Delhi: CSE.

Centro de Defesa dos Direitos Humanos (CDDH) (1993) *Eucalipto-celulose. Desenvolvimento para quem?*, Texeira de Freitas–BA: CDDH.

Centro de Estudos e Pesquisas para o Desenvolvimento do Extremo Sul da Bahia/Centro de Defesa dos Direitos Humanos (CEPEDES/CDDH) (1992) *Eucalipto: uma contradição, impactos ambientais, sociais e econômicos do eucalipto e da celulose no Extremo Sul da Bahia*, Eunápolis-BA: CDDH.

Cerda, Ignacio (1993) 'Aportes para la formulación de política en el sector forestal: el caso chileno, in Alfaro, Marielo et al. (eds.), *Taller Regional: Necesidades y prioridades de investigación en políticas forestales y agroforestales para Latinoamérica*, Costa Rica: IICA.

Cerda, Ignacio, J. Olavarría and M. Abalos (1992) *El sector forestal en Chile: logros y desafíos*, Santiago: INFOR-CORFO, December.

Chalker, Lynda (1992) Letter to Larry Lohmann, 12 November.

Chami Zugman, Isac (1988) 'Reforestation effort pays off well in Brazil', *World Wood*, June.

Chaudhuri, Panya (1995), 'Sowing the Seeds for a New Fiber Supply', *PPI*, March.

Cheney, Karen (1992) 'Fast-Growing Plantations Offer Fiber for US Mills, *World Wood*, October.

Chile (1991) Corporación Nacional Forestal. Memoria anual, Santiago: PAF.

—— (1992a) *Plan de Acción Forestal, Informe Final. Comisión Medio Ambiente*, Santiago: PAF.

—— (1992b) *Plan de Acción Forestal, Informe Final. Comisión Industrias Forestales*, Santiago, PAF.

—— (1992c) *Plan de Acción Forestal, Informe Final. Comisión recursos forestales*, Santiago: PAF.

—— (1992d) *Plan de Acción Forestal. Propuestas nacionales del plan de acción forestal para Chile*, Santiago: PAF.

Churchill, Ward and Jim van der Wall (1988) *Agents of Repression: The FBI's Secret Wars Against the Black Panther Party and the American Indian Movement*, Boston: South End.

Centro Interdisciplinario de Estudios sobre el Desarrollo (CIEDUR) (1991) 'Desarrollo forestal y medio ambiente en Uruguay 16: Relatoria del seminario "Desarrollo forestal: ambiente, economía y sociedad"', Montevideo: CIEDUR.

Clapperton, R. H. (1967) *The Paper-Making Machine: Its Invention, Evolution and Development*, Oxford: Pergamon Press.

Clark, David (1994) 'Fighting for the Future', *Appita Journal* 47, 3: 198–203.

Comité Nacional pro Defensa de la Fauna y Flora (CODEFF) (1991) 'Position paper on the Forest Policy Draft released by the World Bank on April 1, 1991', ms., Santiago: CODEFF.

—— (1992) *El futuro del bosque nativo chileno: un desafío de hoy*, Santiago: CODEFF.

—— (1994) *Hacia un Manejo Sustentable del Bosque Nativo Chileno. Memoria de Taller para Pequeños y Medianos Propietarios*, Santiago: CODEFF.

Colchester, Marcus and Larry Lohmann (eds.) (1993) *The Struggle for Land and the Fate of the Forests*, London and New Jersey: Zed Books.

Coleman, Donald Cuthbert (1958) *The British Paper Industry 1495–1860*, Oxford: Oxford University Press.

Commissão Pastoral da Terra de Minas Gerais (CPT)/Centro Mineiro de Estudos e Pesquisa sobre o Ambiente e Florestas (CEMEPAF)/ Pacific-Asia Resource Center (PARC) (1995) *Pau que nasce certo e entorta a vida dos outros. Monocultura de eucalipto e produção de celulose no Vale do Aço*, Brazil: CTP/CEMEPAF/PARC.

Commonwealth Development Corporation (1992) *CDC: Report and Accounts 1991*, London: CDC.

Compañía Forestal Oriental (1993) *Estudios de impacto ambiental*, Montevideo: Forestal Oriental.

Cooper, K.H. (1990) 'Commercial forestry impact on wetland, grasslands and indigenous forests', in University of Natal, op. cit.

Coordinadora Extremeña de Protección Ambiental (1992) 'Dealing with Disparity: European Structural Funds in South West Spain', *The Ecologist* 22 (3), May/June: 91–96.

Cox, Chris (1993) Presentation at European Rainforest Movement meeting, Oxford, November.

Crawford, Leslie (1991) 'Eucalyptus Joins Chilean Paper Chase', *FT*, 6 June.

Cromer, Robin N. and Glen A. Kile (1993) 'Hardwood Plantation R & D into the 21st Century', *Appita Journal* 46, 4, pp. 293–295.

Cruz, M. Elena and Rigoberto Rivera (1983) *Cambios ecológicos y de poblamiento en el sector forestal chileno*, Santiago: GIA.

—— (1984) *La realidad forestal chilena*, Santiago: GIA.

Dacorso, Daniela (1994) 'Os prós e contras do eucalipto, *Ecología e Desenvolvimento* 3 (44), October: 20–22.

Damiani, Octavio (1990) 'Desarrollo forestal y medio ambiente en Uruguay, 4: Un estudio de caso en el noreste de Canelones', Montevideo: CIEDUR.

Dargavel, John (1991) 'Australia and New Zealand in the Emerging Forest Economy of the Pacific Rim', Proceedings, Australia and New Zealand Institute for Forestry Conference, Christchurch, September–October.

Data Consult, Inc. (1990) *The Pulp and Paper Industry in Indonesia: Its Current State and Prospects*, Jakarta: Data Consult.

Davidson, J. (1987) *Bioenergy Tree Plantations in the Tropics: Ecological Implications and Impacts*, Gland: IUCN.

Deal, Carl (1993) *Greenpeace Guide to Anti-Environmental Organizations*, Berkeley: Odonian.

Dench, Ian (1993) 'Appita's Asia Profile', *Appita Journal* 46, 3: 173–5.

Department of Trade and Industry (1994) *Guide to Gaining Business from World Bank- and Inter-American Development Bank-Funded Projects*, London: DTI.

Dobson, D. (1990) 'Large scale afforestation in Natal/Kwazulu: the challenge facing the small private growers', in University of Natal, op. cit.

Dove, Michael (1992) 'Foresters' Beliefs about Farmers: A Priority for Social Science Research in Social Forestry', *Agroforestry Systems* 17, 13–41.

Down to Earth (DTE) (1991) *Pulping the Rainforest*, London: Down to Earth.

Duarte, Silvana Torquato (1993) *Impactos ambientais do reflorestamentos com eucalipto. A polemica do norte do Espírito Santo e extremo sul da Bahia*, Rio de Janeiro: FASE/IBASE, October.

Dudley, Nigel (1992) *Forests in Trouble: A Review of the Status of Temperate Forests Worldwide*, Gland: WWF.

Dudley, Nigel, Sue Stolton, and Jean-Paul Jeanrenaud (1995) *Pulp Fact: The Environmental and Social Impacts of the Pulp and Paper Industry*, Gland: WWF.

Durning, Alan Thein and Ed Ayres (1994) 'The Story of a Newspaper', *Worldwatch*, November/December, 30–32.

The Ecologist (1993) *Whose Common Future?* London: Earthscan.

Edwards, John C. (1995) 'CEO Forecast: The Recovery Has Arrived', *TAPPI Journal*, January: 30–34.

Edwards, M. B. (1990) 'The Economic Significance of the Forestry and Forest Product Industry to the Natal-Kwazulu region', in University of Natal, op. cit.

Environmental Defense Fund (EDF) (1987) 'Social Forestry in Karnataka', *The Ecologist* 17.

Environmental Justice Networking Forum (EJNF) (1995) 'Report on Paraquat Use in Agriculture in South Africa', ms., Pietermaritzburg.

Equipo de Pastoral Campesina (1993) *Revista Pastoral Campesina* 8, Autumn.

'Eucalyptus Cloning in Veracruz, Bahia, Brazil' (1993) in *Know-How Wire* (Jaakko Pöyry Client Magazine), April.

Evans, Julian (1986) 'Plantation Forestry in the Tropics: Trends and Prospects', *The International Tree Crops Journal* 4.

—— (1992) *Plantation Forestry in the Tropics: Tree Planting for Industrial, Social, Environmental and Agroforestry Purposes*, Oxford: Clarendon Press.

Fábrica Nacional de Papel (1992) 'El dilema del eucalipto', *El Mercado Agropecuario* (Uruguay), 18 June.

Faillace, Sandra and Leandro Valarelli (1994) 'O setor papel e celulose e suas verdes intenções', for.plantation (APC electronic conference), 29 December.

Faillace, Sandra and Moema Miranda (1992) *Relatório Espírito Santo*. Rio de Janeiro: IBASE.

—— (1993) *Dossié Veracruz*, Rio de Janeiro: Federação de Orgãos para Asisténcia Social e Educacional (FASE).

Fairlie, Simon (1992) 'Long Distance, Short Life: Why Big Business Favours Recycling', *The Ecologist* 22 (6), November/December: 276–283.

Fakir, Saliem (1994) Personal communication.

Fanzeres, Anna (1995) Personal communication.

Federação de Orgãos para Asistência Social e Educacional (FASE) (1993) *Inquérito civil sobre fomento florestal*, Vitória: FASE.

Fazzini, Cecilia and Joao Carlos Leal (1988) 'A Souza Cruz planta o futuro, *Senhor* 367, 4 April.

Fearnside, Philip M. (1993) 'Tropical Silvicultural Plantations as a Means of Sequestering Atmospheric Carbon Dioxide', ms., Manaus, 29 June.

Federación Internacional de Trabajadores de la Construcción y la Madera (Federación) (1988) 'Chile: la legislación del trabajo cortada por lo sanó, *Boletín* 3: 12-16.

Ferguson, James (1994) *The Anti-Politics Machine: 'Development', Depoliticization and Bureaucratic Power in Lesotho*, Minneapolis: University of Minnesota.

Fernandez Carro, O. and Wilson, R. (1992) 'Quality Management with Fiber Crops', *TAPPI Journal*, February: 49-52.

Finland National Board of Customs (1990-3) *Statistics*, Helsinki.

Fletcher, Hugh (1988) 'The Pulp and Paper Industry: A New Zealand Perspective', in Schreuder, Gerard F. (ed.) *Global Issues and Outlook in Pulp and Paper*, Seattle: University of Washington.

Floegel, Mark (1994) *The Medium is the Message: Water Pollution, Time Magazine, and Opportunities for Clean Production*, Washington: Greenpeace.

Flynn, Bob (1992) 'Chip Exports Set the Pace for Chile's Forestry Sales', *World Wood*, February.

Food and Agriculture Organization (FAO) (1981) *El eucalipto en la repoblación forestal*, Rome: FAO.

—— (1990) *The Eucalypt Dilemma*, Rome: FAO.

—— (1992) *Mixed and pure plantations in the tropics and subtropics*, Rome: FAO.

—— (1994) *Forest Products Trade Flow Data, Timber Bulletin* vol. XLVII, no. 5, Rome: FAO.

—— (1995) *Forestry Statistics Today for Tomorrow*, Rome: FAO.

Forestal Oriental (1993) 'Estudios de impacto ambiental', Montevideo: Forestal Oriental.

Fourie, J. J. (1990) 'Organized Agriculture's Concern over Largescale Afforestation in Natal/Kwazulu', in University of Natal, op. cit.

Gadgil, Madhav and Guha, Ramachandra (1992) *This Fissured Land: An Ecological History of India*, New Delhi: Oxford University Press.

Galbraith, John Kenneth (1974) *The New Industrial State*, Harmondsworth: Pelican.

Galloway, Leslie (1994) *Closed-Cycle Technologies for Bleached Kraft Pulp Mills*, National Pulp Mills Research Program Technical Report No. 7, Canberra: CSIRO.

Gauthier, J. J. (1991) 'Plantation Wood in International Trade', Rome: FAO.

Geary, Kate (1996) Personal communication.

Geerling, Chris, Dominique N'Sosso, and Gaston Kitemo (1991) *Plan d'Aménagement Environnemental*, Congolaise de Développement Forestier, Pointe Noire: CDF.

George, Susan (1988) *A Fate Worse than Debt*, Harmondsworth: Penguin.

Gómez, Sergio and Echenique, Jorge (1988) 'La agricultura chilena: la dos caras de la modernización', Santiago: FLACSO.

Gonçalves, Marco Antonio (1995) 'Fabrica de papel desencadeia crise no Amapá: Champion adquire fatia da floresta amazônica', *Parabolicas* 2 (12), November.

Gonçalves, Múcio Tosta (1992) 'Aspectos gerais da expanção do refloresta-mento no Brasil' in for.plantation (APC electronic conference), March.

—— (1995) 'Pau que nasce certo e entorta a vida dos outros: Monocultura de eucalipto e produção de celulose no Vale do Aco', Belo Horizonte: Comissão Pastoral da Terra/Centro Mineiro de Estudos e Pesquisa sobre Ambiente e Florestas/Pacific-Asia Resource Center.

Goncalves, Múcio Tosta et al. (1994) *Exploração florestal no norte do ES e sul da Bahia — impactos e alternativas. Estudio de caso: exploração florestal, indústria de celulose e ocupaçao do espa o regional. Relatório final*, Rio de Janeiro: FASE/IBASE, May.

González Bernáldez, Fernando, Manuel Ruiz Pérez, and Catherine Levassor (1989) 'Effets des plantations d'Eucalyptus dans le nord de l'Espagne' Madrid: Universidad Autónoma y Universidad Complutense.

Good, John, Lawson, Gerry and Stevens, Paul (1993) *Natural Environment*, Study No. 8, *Shell/WWF Tree Plantation Review*, London: Shell International Petroleum Company and World Wide Fund for Nature.

Graham, Alastair (1994) 'Wood Flows around the Pacific Rim (A Corporate Picture)', Cygnet, Tasmania: World Wide Fund for Nature.

Grainger, A. (1990) 'Modelling the Impact of Alternative Afforestation Strategies to Reduce Carbon Dioxide Emissions', in Intergovernmental Panel on Climate Change, Response Strategies Working Group (RSWG), Subgroup on Agriculture, Forestry and Other Human Activities, Proceedings of the Conference on Tropical Forestry Response Options to Global Climate Change, Washington: US Environmental Protection Agency.

Grant, Julius, Young, James H., and Watson, Barry G. (1978) *Paper and Board Manufacture*, London: British Paper and Board Industry Federation.

Greenpeace International (1994) *Achieving Zero Dioxin: An Emergency Strategy for Dioxin Elimination*, London: Greenpeace.

Griffin, Rod (1995) 'Taking the Middle Ground from Laboratory to Plantation', *PPI*, August: 28-9.

Groome, Helen (1988) *La Evolución de la Política Forestal en el Estado Español desde el Siglo XIX hasta la Actualidad*, Ph.D. thesis, Autonomous University of Madrid.

—— (1991)'Conflicts Caused by Imbalances in Forest Policy nad Practice in the Basque Country', in *Progress in Rural Policy and Planning* vol. 1, London and New York: Belhaven Press.

Gudynas, Eduardo (1989) 'Amphibians and Reptiles of a Coastal Periurban Ecosystem (Solymar, Uruguay): List, Preliminary Analysis of Community Structure, and Conservation', *Bulletin of the Maryland Herpetological Society* 25 (3) September.

—— (1990) 'La articulación urbano-rural: ecosistemas ecológicos peri-urbanos', *Medio ambiente y urbanización* 8 (31), June.

Gudynas, Eduardo and Rudolf, Juan (1987) 'La herpetofauna de la localidad costera de "Pajas Blancas" (Uruguay): lista sistemática comentada y estructura ecológica de la comunidad', *Com. Mus. Cienc. PUCRS* (Porto Alegre) 46: 173-194.

Guerra, Claudio (1992) *A Case Study in Brazil: Environmental Impacts on the Piracicaba River Basin*, Delft: International Institute for Hydraulic and Environmental Engineering, August.

Guha, Ramachandra (1988) 'Commercial Forestry: Defending the Indefensible', *Deccan Herald*, 19 June.

Hagler, R. (1993) 'Global Forest', *Papermaker*, May: 40-46.

—— (1994) 'The Global Wood Fibre Equation — A New World Order?', in *What is Determining International Competitiveness in the Global Pulp and Paper Industry?* Proceedings, Third International Symposium on Pulp and Paper, Center for International Trade in Forest Products, University of Washington, Seattle, 13-14 September.

—— (1995) 'Outlook for Woodchip Trade in the Pacific Rim', Commodity Markets and Natural Resources: Proceedings of the National Agricultural and Resources Outlook Conference, Canberra, 7-9 February.

Hallowes, David (1995) Personal communication, Environmental Justice Networking Forum, Pietermaritzburg, 18 October.

Hamilton, Clive (1995) 'Woodchipping: Who Pays?', *Wilderness News* (Australia) 140, February/March.

Hamilton, L. S. and King, P. N. (1983) *Tropical Forested Watersheds: Hydrologic and Soils Response to Major Uses or Conversions*, Boulder, Colorado: Westview Press.

Harbinson, Rod (1995) 'Portugal's Eucalyptus Invasion', *SEEDlinks* 17, April.

Harland, Tim (1994) 'Recycled Paper and the Environment', *Permaculture Magazine* 8, 16-18.

Heguy, Eduardo and del Rey, José (coords.) (1981) *Estudios sobre la forestación en el Uruguay*, Montevideo: Univ. de la República.

Helvarg, David (1994) *The War Against the Greens*, San Francisco: Sierra Club.

Higgs, Richard (1992a) 'On the Right Lines', *Paper*, September.

—— (1992b) 'Riocell: Growing Fast', *Paper*, September.

—— (1992c) 'Bahia Debut', *Paper*, September.

—— (1992d) 'Brazil's Japanese Connection', *Paper*, September.

—— (1992e) 'Big and Beautiful', *Paper*, September.

—— (1993a) 'Investing in the Future', *Paper*, August.

—— (1993b) 'Jari: The Mill in the Jungle', *Paper*, August.

—— (1994) 'Battling through Hard Times', *World Paper*, May.

Higham, Robert (1995) *The Paper and Paperboard Industry in Europe and North America: The Future for Producers and End Users*, London: Financial Times.

Hills, R. L. (1988) *Papermaking in Britain 1488-1988*, London: Athlone Press.

Hitchens, Christopher (1994) *For the Sake of Argument*, London: Verso.

Hocking, Jennifer (1991) 'Regulation of Discharge of Organochlorines from Pulp Mills in Canada', *Environmental Management* 15 (2): 195-204.

Holt, J. A. and Spain, A. V. (1986) 'Some Biological and Chemical Changes in a North Queensland Soil following Replacement of Rainforest with *Araucaria cunninghammii* (Coniferae: Araucariaceae)', *Journal of Applied Ecology* 23 (1): 227-237.

Hornby, Donna (1994a) 'The trees are killing us', *The Natal Witness*, 5 December.

—— (1994b) 'Yearning for the return', *The Natal Witness*, 7 December.

Hudson, J. R. (1990) The cane growers' view on timber production, in University of Natal, op. cit.

Hunter, Dard (1978) *Papermaking: The History and Technique of an Ancient Craft*, New York: Dover.

Hutton, Maurice and Winkelmann, Hans (1953) 'Informe sobre forestación y desarrollo forestal en el Uruguay', *Silvicultura* 3 (3), April.

Ikonen, Matti (1994) Personal communication.

Informes y Proyectos SA (INYPSA) (1995) 'Informe ambiental resumen: Fabrica de papel — Transpapel SRL', Fray Bentos: INYPSA.

Inglis, Andrew (1991) 'Travel Report', FAO: Rome, August.

Instituto Brasileiro de Análises Sociais e Econômicas (IBASE) (1993a) 'Cronología sobre a indústria de celulose no Brasil' in for.plantation electronic conference, 3 March.

—— (1993b) 'Cronología sobre a indústria de celulose no Brasil', in for.plantation electronic conference, 6 November.

—— (1994a) 'Cronología sobre a indústria de celulose no Brasil', in for.plantation electronic conference, 3 August.

—— (1994b) 'Cronología sobre a indústria de celulose no Brasil', in for.plantation electronic conference, 10 October.

Instituto del Tercer Mundo (1989) 'Commodities: Why Doesn't the Third World's Salary Go Up?', *Third World Guide 1989-1990*, Montevideo: Editora Tercer Mundo.

Interforest (n.d.) *Interforest Experience in Africa*, Stockholm: Interforest.

—— (n.d.) *Interforest Experience in South and South East Asia*, Stockholm: Interforest.

—— (n.d.) *Lista de Referencias en America Latina*, Stockholm: Interforest.

International Institute for Environment and Development (IIED) (1995) *The Sustainable Paper Cycle: A Report for the Business Council on Sustainable Development* (draft), London: IIED.

Inyaku, Tomoya (1993) 'Why Brazilian pulp is cheap', for.paper (APC electronic conference) 22 April.

Ionides, George N. (1994) 'Technological Developments in Papermaking', presentation to the *Financial Times* Conference on Pulp and Paper, London, 17–18 May.

Jaakko Pöyry (1994) *Address List*, Helsinki: Jaakko Pöyry, 14 November.

—— (n.d. a) *Experience in Asia-Pacific*, Helsinki: Jaakko Pöyry.

—— (n.d. b) *Selected References and Related Experience Regarding Forestry Programming*, Helsinki: Jaakko Pöyry.

—— (n.d. c) *Experience in Sectoral Development of Forestry and Forest Industry*, Helsinki: Jaakko Pöyry.

—— (n.d. d) *Experience in Forestry Project Planning and Implementation*, Helsinki: Jaakko Pöyry.

—— (n.d. e) *Our Commitment to Sustainable Resource Management: Jaakko Pöyry Forest Policy*, Helsinki: Jaakko Pöyry.

Japan International Cooperation Agency (JICA) (1981) *Study Report on the Pulp and Paper Industry Development Program of the Oriental Republic of Uruguay*, Tokyo: JICA.

—— (1987) 'Informe para el estudio del plan maestro para el establecimiento de plantaciones de árboles y utilización de la madera plantada en la República Oriental del Uruguay', Tokyo: JICA.

—— (n.d.) *For Our Green Earth*, Tokyo: JICA.

Japan Paper Association (JPA) (1994) *Pulp and Paper Statistics 1994*, Tokyo: JPA.

Japan Tropical Forest Action Network (JATAN) (1992) *Report on Eucalyptus Plantation Schemes: Investment Activities by Japan's Paper Industry in Brazil and Chile*, Tokyo: JATAN.

—— (1993) 'Report on Eucalyptus Plantation Schemes in Brazil and Chile by Japanese Companies', Tokyo: JATAN

—— (1994) Briefing paper for World Rainforest Movement meeting, New Delhi, April.

Jardine, Kevin (1994) *The Carbon Bomb*, Greenpeace: Amsterdam.

Jawni Odochao (1996) 'Kaan Tham Rai Mun Wian (swidden agriculture)', ms., Chiang Mai: Project for Ecological Recovery.

Jirasak Jindarojana, Juckrit Homchan, Patcharee Saenchan et al. (1993) *Research and Development as Experimental Agriculture with Treated Wastewater. Phase I: Preliminary Experiment and Pre-feasibility Study, Final Report*, Khon Kaen: Khon Kaen University and Phoenix Pulp and Paper Company, Ltd.

Jordan, C. F. (1985) *Nutrient Cycling in Tropical Forest Ecosystems: Principles and their Applications in Management and Conservation*, New York: Wiley.

Jørgensen, Anders Baltzer (1979) 'Forest People in a World of Expansion', *Transactions of the Finnish Anthropological Society*, 2: 77-96.

Joyce, Christopher (1988) 'The Tree that Caused a Riot', *New Scientist*.

Judt, Manfred (1994) 'Hemp: Papermakers Should Take It with a Pinch of Salt', *PPI*, October: 32–35.

Junus, I. (1994) *Indonesia Plywood Industry*, London: WI Carr, Banque Indosuez Group, June.

Kardell, Lars, Eliel Steen, and Antonio Fabiao (1986) 'Eucalyptus in Portugal — A Threat or a Promise?' *Ambio* 15, 1: 6–13.

Kärkkäinen, Matti (1994) Presentation at Jaakko Pöyry Oy, March.

Kerski, Anita (1995) 'Pulp, Paper and Power: How an Industry Reshapes Its Social Environment', *The Ecologist* 25 (4), July/August: 142–9.

Knight, Patrick (1991a) 'Plantations May Support Brazil's First MDF Plant', *World Wood*, April.

—— (1991b) 'Brazil: A Growing Force in World Pulp and Paper', *P&PA*, August.

—— (1993) 'Latin America', *PPI*, July.

Kroesa, Renate (1990) *Greenpeace Guide to Paper*, Vancouver: Greenpeace.

Kuroda, Yoichi (1995) Personal communication, August.

Lagos, Ernesto (1993a) 'No queremos regalos', *Chile Forestal* 202, February.

—— (1993b) 'Proyecto de ley de bosque nativo beneficiará las inversiones', *Chile Forestal* 208, August.

Läitaläinen R. (1992) Letter to Witoon Permpongsacharoen, Project for Ecological Recovery, Bangkok, August.

Lamb, D. (1992) *Exploiting the Tropical Rain Forest*, Paris: UNESCO.

Lang, Chris (1995) Personal communication.

Lara, Antonio (1992) 'Case Study: Chile', in Nigel Dudley (ed.), *Forests in Trouble: A Review of the Status of Temperate Forests Worldwide*, Gland: WWF.

Lara, Antonio and Veblen, Thomas (1993) 'Forest Plantations in Chile: A Successful Model', in Alexander Mather (ed.), *Afforestation: Policies, Planning and Progress*, London: Belhaven Press.

Larrobla, Ricardo et al. (coords.) (1992) *Plantemos árboles. Guía práctica para el forestador*, Montevideo: Rotary Club/Hemisferio Sur.

Latorre, Jaime (1991) 'Monocultivo forestal: mito y realidad', *Uruguay Forestal* 1, July.

Latorre, Raúl (1995) 'Informe sobre empleo y capacitación en la forestación', ms.

Le Roux, S. D. (1990) 'Impact of Afforestation on the Agricultural Resources of Natal', in University of Natal, op. cit.

Letto, Jay (1994) 'Go East, Young Timberman!' *E Magazine*, Jan./Feb.: 26-33.

Leyton, José et al. (1986) 'El fomento de la actividad forestal y su impacto sobre el desarrollo rural en Chile', in *El Desarrollo Frutícola y Forestal en Chile y Sus Derivaciones Sociales*, Santiago: CEPAL.

Ligrone, Atilio (1992) 'Situación actual y política del fomento del sector forestal nacional', in *Plantemos árboles. Guía práctica para el forestador*, Montevideo: Rotary Club.

Lima, Walter de Paula (1993) *Impacto Ambiental do Eucalipto*, São Paulo: EDUSP.

Lohmann, Larry (1991) 'Peasants, Plantations and Pulp', *Bulletin of Concerned Asian Scholars* 23, 4: 3-17.

—— (1993) 'Green Orientalism', *The Ecologist* 23, 6.

—— (forthcoming a) 'No Rules of Engagement: the Creative Politics of "Environment" in Thailand', in J. Rigg (ed.) *Counting the Costs: Economic Growth and Environmental Change in Southeast Asia*, Singapore: ISEAS.

—— (forthcoming b) 'Freedom to Plant: Indonesia and Thailand in a Globalizing Pulp and Paper Industry' in Michael J. G. Parnwell and Raymond Bryant (eds.), *Environmental Change in South-East Asia: Rendering the Human Impact Sustainable*, London: Routledge.

Lohmann, Larry, Thomas Wallgren and Witoon Permpongsacharoen (forthcoming) 'Intercultural Politics: the Coevolution of a Master Plan and Its Critics', in Gudeman, S. and Marglin, S. (eds.) *People's Ecology, People's Economy*.

López, Elvio and Cussac, Carlos (1943) *Arboles forestales en el Uruguay y problemas afines*. Montevideo: Mercant.

Lugo, A.E., Brown, S. and Chapman, J. (1988) 'An Analytical Review of Production Rates and Stemwood Biomass of Tropical Forest Plantations', *Forest Ecology Management* 23: 179-200.

Luukkanen, Olavi and Hakulinen, Mervi (eds.) (1991) *From Bangkok to the Blue Nile: Review of the First Decade of the Tropical Silviculture Research Group 1980-1990 and Abstracts of Research Reports*, Tropical Forestry Reports 7, Helsinki: University of Helsinki Department of Silviculture.

MacIsaac, Ronald and Champagne, Anne (1994) *Clayoquot Mass Trials: Defending the Rainforest*, Gabriola Island, B.C.: New Society Publishers.

Magee, Sean (1994) Letter to Chris Lang, Earth Action Resource Centre, 1 December.

Malee Traisawasdichai (1994) 'Laotians "Look Likely Losers" from Foreign-Funded Forestry', *The Nation* (Bangkok), 11 July.

Marchak, M. Patricia (1991) 'For Whom the Tree Falls: Restructuring of the Global Forest Industry' in *BC Studies* 90, Summer: 3-24.

—— (1992) 'Latin America and the Creation of a Global Forest Industry', in H. K. Steen and R. P. Tucker (eds.), *Changing Tropical Forests: Historical Perspectives on Today's Challenges in Central and South America*, New York: Forest History Society, International Union of Forestry Research Organizations.

Marcus, Amanda (1993a) 'Klabin Enjoys Lead Role among Latins', *PPI*, November.

—— (1993b) 'CELPAV Paves the Way for More Woodfrees', *PPI*, December.

Márquez Fernández, Dominga (1985) 'Las repoblaciones de eucaliptos y su impacto en la última decada 1973-83', *Revista de Estudios Andaluces* 5.

Martone, Francesco (1993) 'Brazil: Anti-Pulp Activist Killed', in for.paper electronic conference, 3 May.

—— (1995) *Overview on Current and Projected Trends in Trade, Production and Consumption of Pulp and Paper Products; Details on Main Paper TNCs; Trade Analysis for Selected Countries*, Rome: Greenpeace International.

Masaki, Mikio (1995) 'Commercial Reforestation and Japanese ODA', ms.

Mather, Alexander (1990) *Global Forest Resources*, London: Belhaven.

Matías, Moisés (1994) 'A implantação da indústria de celulose no Maranhão: CELMAR na região Tocantiná', Pará: Sociedade Paraense de Defesa dos Direitos Humanos.

McClelland, W. Craig (1994) 'Issues for the 1990s: US Pulp and Paper Industry', presentation at *Financial Times* conference on World Pulp and Paper, London, 17–18 May.

Megalli, Mark and Friedman, Mark (1991) *Masks of Deception: Corporate Front Groups in America*, Washington: Essential Information.

Messner, Dirk (1993) 'Búsqueda de competitividad en la industria maderera chilena', *Revista de la CEPAL* 49, April.

Mikami, Takaito (1991) 'Forestación: base de la construcción de una nación', Montevideo: JICA.

Miranda, Moema (1992) 'O extremo sul da Bahia e a avassaladora chegada da modernidade. II. relatorio de viagem ao extremo sul da Bahia', Rio de Janeiro: FASE/IBASE, October.

—— (1993a) 'Papel e celulose no Brasil: novos projetos, velhos dilemas', Rio de Janeiro: IBASE.

—— (1993b) 'Nas pontas do extremo: outras historias, outros tempos', Rio de Janeiro: FASE/IBASE, April.

—— (1993c) 'Indios reivindicam suas terras', for.plantation (APC electronic conference) 14 July.

Moniaga, Sandra (1993) 'CIFOR Establishment: Consistent with Conservation?', unpublished paper, Pontianak, Kalimantan.

Morey, Carlos and Juan Porcile (1992) 'Protección sanitaria de los bosques: necesidad de una estrategia nacional', in *Plantemos árboles. Guía práctica para el forestador*, Montevideo: Rotary Club/Hemisferio Sur.

Morrison, Elaine and Stephen Bass (1992) 'What about the People?' in Sargent and Bass (eds.), op. cit.

Moulton, R. J. and K. R. Richards (1990) *Costs of Sequestering Carbon through Tree Planting and Forest Management in the United States*, Washington: US Department of Agriculture Forest Service.

Muzio, Gabriele (1995) 'A Contribution to the Discussion on Forest Plantations', ms., March.

Muzio, Gabriele and Rosario Ortiz (1995a) 'A Brief Commentary on the FAO's Draft Report on "Combating Deforestation"', unpublished ms.

—— (1995b) 'Another FAO-inspired "Green Revolution"', unpublished.

Myers, Norman (1989) *Deforestation Rates in Tropical Forests and their Climatic Implications*, London: Friends of the Earth.

—— (1991) Presentation at Issue Dialogue on 'Tree Plantations: Benefits and Drawbacks', Centre for Applied Studies in International Negotiations, Geneva, 23–24 April.

—— and Goreau, T. J. (1991) 'Tropical Forests and the Greenhouse Effect: A Management Response', *Climate Change* 19 (1–2): 215–225.

Native Forest Network (NFN) (1995) 'Pulp in the Machine: International Day of Action against Pulp and Paper', *Earth First!* Beltane.

Nectoux, Francois and Yoichi Kuroda (1989) *Timber from the South Seas: An Analysis of Japan's Tropical Timber Trade and Its Environmental Impact*, Gland: WWF.

Nelson, Joyce (1994) 'Pulp and Propaganda', *Canadian Forum*, Jul./Aug.: 14–20.

Niku, Pekka (1993) 'Worldwide Review of Recycled Fibre', *Know-How Wire* (Jaakko Pöyry Client Magazine) 1: 6–7.

Oberlander, Ronald Y. (1994) 'A Canadian Perspective: Uncommon Approaches to Common Problems', paper presented to the Financial Times Conference on Pulp and Paper, London, 17–18 May.

O'Brien, Mary (1990) 'A Crucial Matter of Cumulative Impacts: Toxicity Equivalency Factors', *Journal of Pesticide Reform*, Summer.

Oinn, David (1994) Presentation at PIRA teach-in on paper, Leatherhead, Surrey, 7 December.

Olsson, Roger (ed.) (1995) *The Taiga Trade: A Report on the Production, Consumption and Trade of Boreal Wood Products*, Jokkmokk: Taiga Rescue Network.

O'Neill, John (1993) *Ecology, Policy and Politics: Human Well-Being and the Natural World*, London: Routledge.

Ong'wen, Oduor (1994), 'Death Mill: A Case Study of Pan African Paper Mills (EA) Ltd., Webuye, Kenya', prepared for the NGO Symposium during the Annual Meeting of the Board of Governors of the African Development Bank, Nairobi, May.

Organization of American States (1994) *Uruguay: proyecto regional de alternativas para la inversión forestal*, Washington: OAS.

Orton, David (1994) Personal communication.

Otero, Luis (1990) 'Impacto de la actividad forestal en comunidades locales en la VIII Región', *Ambiente y Desarrollo* 6 (2), August: 61–69.

Overseas Development Administration (1993) *British Overseas Aid Annual Review 1992*, London, ODA.

Oxfam (1994) 'Fifty Years is Enough', Oxford: Oxfam.

Ozinga, Saskia (1994) 'Production and Consumption of Timber and Timber Products', Amsterdam: Milieu Defensie.

Paavilainen, Leena (1993) 'Non-wood Plants: A Potential Source of Fibre', in *Know-How Wire* (Jaakko Pöyry Client Magazine), 2: 8–9.

Panario, Daniel et al. (1991) *Desarrollo forestal y medio ambiente en Uruguay 11: Hacia una evaluación de efecto ambientales de la forestación con especies introducidas*, Montevideo: CIEDUR.

Pandey, Devendra (1992) *Assessment of Tropical Forest Plantation Resource* (draft), Umeå: Swedish University of Agricultural Sciences.

Paper Europe (1995) Conference Programme for Third Global Conference on Paper and the Environment, London, 27–28 March.

Paper Publications (1994) *UK Directory of Papers and Boards*, London: Paper Publications.

Pappens, Rita Anna (1995) 'Market Pulp: From Bust to Boom . . . and Back Again', *PPI*, January: 76.

Paré, Luisa (1992) 'El eucalipto y la integración de México al mercado internacional', *El Cotidiano* 48, June.

Partido Socialista Obrero Español (PSOE) (1979) *El eucalipto: Actas de las jornadas de trabajo sobre el eucalipto*, Huelva: PSOE.

Pasuk Pongpaichit (1995) Presentation at Seminar on Community Rights and Environmental Conservation: The Case of the Paper Pulp Industry, Chulalongkorn University, Bangkok, 27 April.

Payer, Cheryl (1991) *Lent and Lost*, Pluto: London.

Penna, Ian (1992) *Japan's Paper Industry: An Overview of Its Structure and Market Trends*, Tokyo: Friends of the Earth Japan.

—— (1994) 'Daishowa Paper Manufacturing Co.: A Corporate Profile', Tokyo: JATAN.

Perez Arrarte, Carlos (comp.) (1993) *Desarrollo forestal y medio ambiente*, Montevideo: CIEDUR/Hemisferio Sur.

—— (1995) 'Impactos socioeconómicos', in Ricardo Carrere et al., *Impactos de la Forestación en Uruguay*, Montevideo: Instituto del Tercer Mundo/REDES.

Pesonen, Kari (1995) 'Indo-China P & P Industry — Its Future Prospects', *Paper Asia*, May: 11–18.

Pollett, E. A. and M. Mander (1995) *Mosa Florestal Afforestation Project, Moçambique: Report on Preliminary Field Visits and Scoping Workshops Held at Maputo, Zitundo and Ponta do Ouro to Identify Potential Environmental Issues*, Pietermaritzburg: Institute of Natural Resources, January.

Poore, D. and C. Fries (1985) *Efectos ecológicos de los eucaliptos*, Rome: FAO.

Porter, R.N. (1990) 'Future Afforestation and the Potential Impacts on Nature Conservation in Natal', in University of Natal, op. cit.

Posey, Darrell (1985) 'Indigenous Management of Tropical Forest Ecosystems: The Case of the Kayapó Indians of the Brazilian Amazon', *Agroforestry Systems* 3: 139–158.

—— (1990) 'Cultivating the Forests of the Amazon: Science of the Mebengokre', *Orion Nature Quarterly* 9 (3): 16–23.

Postel, Sandra and L. Heise (1988) *Reforesting the Earth* (Worldwatch Paper No. 83), Washington: Worldwatch Institute.

Pou, Rosario (1991) 'Algo más sobre el eucalipto' *Uruguay Forestal*, July.

—— (1992) 'Tratamientos silviculturales', in *Plantemos árboles. Guía práctica para el forestador*, Montevideo: Rotary Club.

Pratt, Larry and Ian Urquhart (1994) *The Last Great Forest: Japanese Multinationals and Alberta's Northern Forests*, Edmonton: NeWest Press.

Price, David (1995) 'Europe's Disposable Paper Market: The Going Gets Tough', *P&PA* 21, May.

Project for Rural Ecological Development (PRED) (1996) Seminar on Research and Planning on Eucalyptus, Kaset Wisai district, Roi Et.

Rackham, Oliver (1986) *The History of the Countryside*, London: Dent.

Rada, José Antonio (1992) 'Plantaciones de eucalipto en Chile', *Chile Forestal* 196, July.

Rahikainen, Anne, Dorothy Paun and Lynn Catlett (1995) 'An Examination of the United States Pulp and Paper Industry', CINTRAFOR Working Paper 54, College of Forest Resources, Seattle.

Rajesh, Noel (1995) 'Thailand's Pulp and Paper Industry: An Overview', Bangkok: Project for Ecological Recovery.

Rasmusson, Ulf (1994) *Swedish/Scandinavian Involvement in Indonesian Forestry — The Industrial Forest Plantation and Pulp Mill Sector*, Stockholm: World Wide Fund for Nature.

Rice, Tim (ed.) (1995) *Out of the Woods: Reducing Wood Consumption to Save the World's Forests: A Plan for Action in the UK*, London: Friends of the Earth.

Rich, Bruce (1994) *Mortgaging the Earth: The World Bank, Environmental Impoverishment, and the Crisis of Development*, Beacon: Boston.

Riddlestone, Sue, Pooran Desai, Tim Rice and Tom Solly (1995) 'The Reclamation, Reuse and Recycling of Post-Consumer Primary Industrial Wood Product Waste' in Tim Rice (ed.) *Out of the Woods: Reducing Wood Consumption to Save the World's Forests: A Plan for Action in the UK*, London: Friends of the Earth.

Rocky Mountain Institute (1990) 'A $20 Bill on the Sidewalk', *RMI Newsletter*.

Rogers, Lewis (1955) 'Informe sobre forestación en el Uruguay', *Silvicultura* 8 (31), May.

Rosoman, Grant (1994), *The Plantation Effect: An Ecoforestry Review on the Environmental Effects of Exotic Monoculture Tree Plantations in Aotearoa/New Zealand*, Wellington: Greenpeace.

Ruiz Pérez, Manuel (1990) 'El eucalipto: un tigre de papel?' unpublished report prepared for Greenpeace Spain.

Ryan, Megan (1994) 'Recycling on a Roll' in WorldWatch Institute, *Vital Signs 1994*, New York: Norton.

Said, Edward (1978) *Orientalism*, New York: Pantheon.

Samaja Parivartana Samudaya (SPS) (1989) *Whither Common Lands?* Dharwad: SPS.

Sandbrook, Richard (1995) Personal communication.

Sanitsuda Ekachai (1992) 'Man and the Forest', *Bangkok Post*, 24 January.

Sanways, M. J.and Moore, S. D. (1991) 'Influence of Exotic Conifer Patches on Grasshopper (Orthoptera) Assemblages in a Grassland Matrix at a Recreational Reserve, Natal, South Africa', *Biological Conservation* 57: 117–137.

Sargent, Caroline (1990a) *The Khun Song Plantation Project*, London: International Institute for Environment and Development.

—— (1990b) Personal communication.

—— (1992) 'Natural Forest or Plantation?' in Sargent and Bass (eds.), op. cit.

Sargent, Caroline and Stephen Bass (eds.) (1992) *Plantation Politics: Forest Plantations in Development*, London: Earthscan.

Sargent, Caroline, Mingsarn Khaosa-ard, John Butler and Pieter van Ginneken (1992) *Thai Forestry Sector Master Plan: Mid-Term Review*, London: International Institute for Environment and Development.

Sawyer, Jacqueline (1993) *Plantations in the Tropics: Environmental Concerns*, Gland: International Union for the Conservation of Nature.

Saxena, Naresh C. (1992a) 'Farm Forestry and Land-use in India: Some Policy Issues', *Ambio* 21 (6), 420–5.

—— (1992b) 'India's Eucalyptus Craze: The God that Failed', *Panoscope*, May.

Schindler, David (1995) 'Toward Sustainable Development of the Canadian Pulp and Paper Industry', in Western Canada Wilderness Committee, *Al-Pac: Mitsubishi's Attack on Alberta*, Edmonton: WCWC.

Schinke, Gert (1992a) 'Brazil pulp mill scandal', in haz.pulpmills (APC electronic conference), 10 June.

—— (1992b) 'Alerto: Riocell escandalo', in haz.pulpmills (APC electronic conference), 12 June.

Schlatter, Roberto and Murúa, Roberto (1992) 'Control biológico de plagas forestales: bosque artificial y biodiversidad', *Ambiente y Desarrollo* 8 (1), April: 66–70.

Schreuder, Gerard F. and E. Anderson (1988) 'International Wood Chip Trade: Past Developments and Future Trends, with Emphasis on Japan', in Gerard F. Schreuder (ed.) *Global Issues and Outlook in Pulp and Paper*, Seattle: University of Washington.

Scott, James C. (1990) *The Arts of Resistance: Hidden Transcripts*, New Haven: Yale University Press.

Shell International Petroleum Company and World Wide Fund for Nature (1993) *Tree Plantation Review*, 11 vols., London: Shell/WWF.

Shiva, Vandana (1991a) *The Violence of the Green Revolution*, Penang: Third World Network and London and New Jersey: Zed Books.

—— (1991b) *Ecology and the Politics of Survival: Conflicts over Natural Resources in India*, New Delhi: Sage.

—— (1993) *Monocultures of the Mind: Perspectives on Biodiversity and Biotechnology*, Penang: Third World Network and London and New Jersey: Zed Books.

Shiva, Vandana and J. Bandyopadhyay (1987) *Ecological Audit of Eucalyptus Cultivation*, Dehra Dun: Research Foundation for Science and Ecology.

Shiva, Vandana, H. C. Sharatchandra and J. Bandyopadhyay (1982) 'Social Forestry: No Solution within the Market', *The Ecologist* 12.

SKEPHI (NGO Network for Forest Conservation in Indonesia) (1995) 'The Development Impacts of Industrial Planting Forest (HTI) Belongs Barito Pacific Group Company and South Sumatera', ms., Jakarta.

Smith, Maureen (1995) 'The Paper Industry: Agenda for Reform' in Soltani, Atossa, op. cit.

—— (forthcoming) *The Paper Industry and Sustainable Production: An Environmental Argument for Industrial Restructuring*, Cambridge, MA: Massachusetts Institute of Technology Press.

Soetikno, Abubakar (1993) 'Appita's Asia Profile', *Appita Journal* 46, 2: 90–92.

Soltani, Atossa and Whitney, Penelope (eds.) (1995) *Cut Waste, Not Trees: How to Save Forests, Cut Pollution and Create Jobs*, San Francisco: Rainforest Action Network.

Sonnenfeld, David (1995) Personal communication.

Soulas, Alain (1994) 'Post-Recession Global Strategies: Competition, Trade and the Environment', paper presented to the Financial Times Conference on World Pulp and Paper, London, 17–18 May.

Spencer, J. E. (1966) *Shifting Cultivation in Southeast Asia*, Berkeley: University of California.

Stauber, John C. and Sheldon Rampton (1995) *Toxic Sludge is Good for You: Lies, Damn Lies and the Public Relations Industry*, Monroe, ME: Common Courage.

Stefan, Virginia (1995) 'Market Pulp Figures Signal a New Brand of Recovery', *PPI*, August: 14–16.

Stolovich, Luis (1995) *Forestación ¿Un Negocio Para Quién?* Montevideo: Instituto del Tercer Mundo/CIEDUR.

Suppachai Saranaraksa (1996) Interview, Phoenix Pulp and Paper Co., Ltd., 21 February.

Susaeta, Eladio (1993) 'Madera pulpable: interesante opción para el Uruguay', *Uruguay Forestal* 5, May.

Swann, Charles (1992) 'Mexico: Importer of pulp and paper', *Papermaker*, November.

—— (1993) 'South America: The Promise and the Problems', *Papermaker*, August.

Taiga Rescue Network (TRN) (1993) 'The Taiga: A Treasure or Timber and Trash?' Jokkmokk: TRN.

Terra, Patrícia (1992) 'Papel: uma indústria poluente', *Ecologia e Desenvolvimento* 2 (19), September: 18–20.

Thailand Royal Forest Department (RFD), Ministry of Agriculture and Cooperatives, Royal Thai Government (1993), *Thai Forestry Sector Master Plan*, Bangkok: RFD.

Thomen, Antonio (1990) Personal communication.

—— (1991) 'Desierto verde: la conspiración del eucalipto', *Ecología y Conservacionismo* 15, May.

Tjahjono, S. Indro (1993) 'The Tragedy of Indonesian Forests', ms., Jakarta.

Tunya Sukpanich (1990) 'Killing the Land for Money', *Bangkok Post*, 22 February.

Tuset, Rinaldo et al. (1990) *El Complejo Forestal en Uruguay*, Montevideo: Fac. Agronomía.

Ulvila, Marko (1994) 'Pulp Industry in Indonesia: Finnish Companies Participate in Rainforest Destruction', Tampere: Coalition for Environment and Development.

United Nations (1993-4), *Import-Export Statistics*, New York: UN.

United States Department of Agriculture (USDA) Forest Agriculture Service (1994) *Wood Products Trade and Foreign Markets*, Washington: USDA, August.

University of Natal Institute of Natural Resources (1990) *The Physical, Social and Economic Impacts of Large-scale Afforestation in Natal/Kwazulu*, Proceedings of the Forestry Impacts Workshop, Pietermaritzburg: Institute of Natural Resources, May.

Uruguay Ministerio de Ganadería, Agricultura y Pesca (MGAP) (1994a) *Forestación: negocios de buena madera*, Montevideo: MGAP.

—— (1994b) *Uruguay forestal: más naturaleza. Oportunidades de inversion*, Montevideo: MGAP.

Usher, Ann Danaiya (1990a) 'A Forest Policy Sadly Gone Awry' in *The Nation* (Bangkok), 10 May.

—— (1990b) 'Eucalyptus – Widening the Gap', in *The Nation* (Bangkok), 14 June.

—— (1991) 'A Finn-Ancial Harvest', *The Nation* (Bangkok), 10 February.

—— (1994) 'Swedes Assisting FIO Reform', *The Nation* (Bangkok), 13 December.

—— (1994) Personal communication, December.

Vail, David (1993) 'The Internal Conflict: Contract Logging, Chainsaws and Clear-cuts in Maine Forestry' in Tariq Banuri and Frédérique Apffel-Marglin (eds.) *Who Will Save the Forests? Knowledge, Power and Environmental Destruction*, London and New Jersey: Zed Books.

Valarelli, Leandro Lamas (1992) 'Complexos Florestais de Celulose: O Mito da Modernidade', *Proposta* 53, May.

Vandermeer, John and Ivette Perfecto (1995) *Breakfast of Biodiversity: The Truth about Rain Forest Destruction*, Oakland: Food First.

van der Zel, D. W. (1985) *A Decision-Making Modelling Approach to Mountain Catchments Management in Southern Africa*, Unpublished Ph.D. thesis, University of Natal.

—— (1990) 'The Past, Present and Future of the Afforestation Permit System in the R.S.A.', in University of Natal, op. cit.

van Dijk, Martha and Ariette Dekker (1995) 'Industry Faces Another Cycle When Tapping Finance Markets', *PPI*, September: 23–27.

van Ginneken, Pieter (1991) *Not Seeing the People for the Trees? A Review of Practices and Policies of Tree Planting*, Amsterdam: AIDEnvironment.

van Hauwermeiren, Saar (1995) 'La sustentabilidad del ciclo de papel: aspectos criticos para Chile', ms., Santiago: Instituto de Ecologia Politica.

van Hook, Matthew (1994) Paper presented to the Financial Times Conference on Pulp and Paper, London, 17–18 May.

Wadsworth, John (1993) 'Perspectiva de los mercados y comercialización en el marco del desarrollo de las plantaciones de eucalipto en Uruguay', *Uruguay Forestal* 6, November.

Wahana Lingkungan Hidup Indonesia (WALHI) (1995) Letter to OECF, Jakarta, 14 July.

—— and Yayasan Lembaga Bantuan Hukum Indonesia (YLBHI) (1992) *Mistaking Plantations for Indonesia's Tropical Forest*, Jakarta: WALHI and YLBHI.

Wallgren, Thomas (1994) Personal communication.

Wannasri Boonyarattaphan (1994) *Phonkaansamruaj lae wikhroh phonkrathope buengton khrongkaan kaanchainamsia phueakaankaset borisat phoenix pulp and paper jamkat* (Preliminary results of survey and analysis of Project Green, Phoenix Pulp and Paper, Ltd.), Khon Kaen, Thailand: Khon Kaen University.

Wapenhans, Willi (1992) *Effective Implementation: Key to Development Impact*, confidential discussion draft, World Bank, Washington: World Bank, 24 July.

Western, A. W. (1979) *Small-Scale Papermaking*, Rugby: Intermediate Technology Information Services.

Western Canada Wilderness Committee (WCWC) (1994) *Al-Pac: Mitsubishi's Attack on Alberta*, Edmonton: WCWC.

Westoby, Jack (1987) *The Purpose of Forests*, Oxford: Blackwell.

—— (1989) *Introduction to World Forestry*, Oxford: Blackwell.

Weyerhaeuser (1995) *Weyerhaeuser Report 1994*, Tacoma.

Whitham, Jeremy (1994) 'The Foreign Expansion of the Japanese Pulp and Paper Industry', in *What is Determining International Competitiveness in the Global Pulp and Paper Industry?* Proceedings, Third International Symposium on Pulp and Paper, Center for International Trade in Forest Products, University of Washington, Seattle, 13–14 September.

Wicht, C. L. (1967) 'Forest Hydrology Research in the South African Republic', in *Forest Hydrology*, New York: Pergamon Press.

Willums, Jan-Olaf and Ulrich Goluke (1992) *From Ideas to Action: Business and Sustainable Development*, Oslo: ICC.

Wilson, Lisa (1994) 'No Sex Please, We're Clones', *New Scientist* 24/31 December: 26–28.

Wilson, Philip (1995) 'Getting Back to the Roots of Tree Improvement', *PPI*, April

Wilson, Robert A. (1991) 'Managed Forests: Economic and Ecological Aspects', presentation to the World Pulp and Paper Environmental Conference, Leningrad, 16–17 April.

—— (1995) 'Learning Lessons from Intelligent Fibre', *PPI*, April: 45-6.

Wong, Alfred (1991) 'Industry Development and Environmental Protection — Compatible Goals?' *Proceedings, Tropical Pulp Conference*, Jakarta: Arbokem Inc. and Deutsche Papierwirtschaft.

—— (1992) 'New Direction in Industry Development and Environmental Protection for Non-Wood Pulp Mills in Developing Countries', ms., Vancouver: Arbokem Inc.

——, Romuald Krzywanski and Chen Chiu (1993) 'Agriculture-Based Pulp as Secondary Papermaking Fibres', *TAPPI Proceedings*, Pulping Conference.

World Bank (1989) *Uruguay Second Agriculture Development Project*, Staff Appraisal Report, Washington: World Bank.

—— (1994a) *Indonesia Environment and Development: Challenges for the Future*, Washington: World Bank.

—— (1994b) *Forest Policy Implementation Review* (draft), Washington: World Bank.

—— (1994c) *China Forest Resource Development and Protection Project*, Staff Appraisal Report, Washington: World Bank.

—— (1995a) *World Development Report 1994*, Washington: World Bank.

—— (1995b) *Monthly Operational Summary of Bank and IDA Proposed Projects*, November.

—— (n.d.) *Guide to International Business Opportunities*, Washington: World Bank.

World Resources Institute (1994) *World Resources 1994-95*, New York: Oxford University Press.

Worldwatch, *Vital Signs* 1994, New York: Norton.

Wright, Roger (1993) 'World Pulp Market: Forecasts and Prospects as at Mid-1992', *P&PA* 14, 13–19.

—— (1994) 'New Markets; New Developments – Indonesia', presentation to the *Financial Times* Conference on World Pulp and Paper, London, 17-18 May.

Wright, Roger (1995) Personal communication.

Zerner, Charles (1992) *Indigenous Forest-Dwelling Communities in Indonesia's Outer Islands: Livelihood, Rights, and Environmental Management Institutions in the Era of Industrial Forest Exploitation*, unpublished report commissioned by the World Bank.

Periodicals

Asia Pulp and Paper (AP&P)
Bangkok Post (BP)
Chile Forestal
Deccan Herald
Development Today (DT) (Oslo)
Dirigente Industrial
Down to Earth (DTE) (Delhi)
Down to Earth (DTE) (London)
The Economist
Environesia (Jakarta)
Environmental Science and Technology (ES&T)
Enso Vision
Far Eastern Economic Review (FEER)
Financial Times (London)
Forum Keadilan (Jakarta)
The Guardian (G) (London)
Helsingen Sanomat (Helsinki)
The Independent (London)
International Agricultural Development (IAD)
International Herald Tribune (IHT)
Jagruta Vani (JV), quarterly newsletter of Samaja Parivartana Samudaya (Dharwad, Karnataka)
Jakarta Post (JP)
Japan Pulp and Paper (JP&P)
Know-How Wire (Jaakko Pöyry Client Magazine)
Kompas (Jakarta)
Krungthep Thurakit (Bangkok)
Management and Technology (MT) (Leatherhead, UK)

Matichon (Bangkok)
The Nation (Bangkok)
New Straits Times (NST) (Singapore)
Observer (London)
Paper
Paper and Packaging Analyst (P&PA)
Paper Asia
Phuu Jatkaan (Northeastern edition, Khon Kaen, Thailand)
Phuu Jatkaan Raai Sapdaa (Bangkok)
Phuu Jatkaan Raai Wan (Bangkok)
Public Relations Watch (PRW) (Madison, WI, USA)
Pulp and Paper (P&P)
Pulp and Paper International (PPI)
Paperboard Packaging (PP)
Papermaker
Prachachart Thurakit (Bangkok)
La República (Montevideo)
Revista Pastoral Campesina
Setiakawan (Jakarta)
Skogsindustrierna Press Releases
Statesman (New Delhi)
Suara Pembuaran (Jakarta)
Taiga News (TN)
TAPPI Journal
Tempo (Jakarta)
The Times (London)
Treasury News (Washington, DC)
Uruguay Forestal
World Rainforest Report (WRR) (Lismore, Australia)
World Wood (WW)

Index

acacia, 3, 6, 7, 8, 56, 58, 72, 134, 198, 207, 217, 220, 221, 222, 230, 239, 241
Acos Espeçiais Itabira, 159, 160, 161
Adlard, Philip G., 141
afforestation, 14, 45, 68, 83, 150, 156, 190–91, 197–98, 205–06, 209
Afforestation Permit System, 205–06
Ahlstrom, 53, 91, 211, 215
Algeria, 97
Allende, Salvador, 171
Alvim, Paulo de T., 164
Amapá Florestal e Celulose S. A., 163
Amcor, 46, 112–13
Andinos, 175
Angelini, Anacleto, 176
Angola, 7, 9
Aracruz, 62, 88, 103, 109, 149–55, 159, 160, 167
Argentina, 5, 9, 14, 38, 44, 50, 57, 73, 101, 109, 131, 149
Arjo Wiggins Appleton, 51, 54, 88, 110–11, 121, 149
Arnold, Bruce, 50
Asea Brown Boveri (ABB), 91, 122, 211, 215
Asia Pacific Resources International, 217
Asia Pulp & Paper, 217
Asia Tech, 230, 232, 239, 241
Asian Development Bank, 90, 97, 134, 235, 240
Aspex Paper, 213
Associação Brasileira de Exportadores de Celulose (ABECEL), 92
Astillas Exportaciones, 171
Astra, 228, 242
Australia, 3, 5, 9, 15, 18, 38, 41, 42, 43, 45, 46, 48, 50, 54–58, 63, 88, 89, 103, 104, 109, 111–13, 120, 198, 214, 234, 235, 241
Austria, 44, 91, 99, 107, 109, 112, 200, 215
Avenor, 31, 33, 110

Bahia Sul, 62, 88, 104, 149, 155–58, 160
Ballarpur, 46, 230
bamboo, 13, 16, 18, 28, 239, 241
Bangladesh, 7, 38, 97, 214
Barclays Bank, 230
Barito Pacific International, 212, 215–17, 220, 223, 226, 228, 252; Tanjung Enim Lestari (TEL) installation, 88, 212, 216, 217, 220, 223, 227, 228, 252
Bass, Stephen, 151
Batak, 225
Bazett, Michael D., 6, 8, 40, 49
Belgium, 44, 47, 215

Belgo-Mineira Company, 159–61
Beloit, 91, 92, 215–16, 229
Bermuda, 217
Bethlehem, Lael, 201
Bhutan, 97
bilateral 'aid' agencies, 12, 89, 93–95, 100, 137, 143, 159, 192, 216–17, 219, 242, 252
Billerud, 150
biodiversity, industrial plantations' threat to, 45, 51, 63, 69–75, 122, 131–32, 142, 153, 158, 161, 163–64, 168, 180–82, 193, 196, 207–09, 219–23, 235–36, 249
Birla, 237
Black Clawson, 91
bleach, 20, 23, 81, 149, 154–55, 160, 165–67
BNDES (National Economic and Social Development Bank, Brazil), 148, 150, 153, 155, 162, 167, 168
Board for Investment and Technical Support (BITS), 96, 234
Bolivia, 7, 38, 137
Borregaard, 165
Bowater, 110–11
Brazil, 4, 6–7, 8, 9, 12, 21, 22, 24, 26, 29, 32, 35, 38, 39, 41–47, 49, 50, 52, 53, 59, 60, 62, 65, 70, 71–72, 73, 74, 78, 82, 83, 86, 88, 89, 97, 101, 103, 107, 109, 111–3, 116, 126, 131, 135, 137, 147–68, 237
bribes, 115, 117, 121, 166, 225, 236
Brierly Investments, 88, 176, 230
British-American Tobacco, 88, 150
Bulgaria, 38
Burapha Development Consultants, 90
Burkina Faso, 7
Burma, 241
Burson-Marsteller, 123–24
Burundi, 7, 8, 49
Business Council for Sustainable Development, 151

Cal, Roberto, 193
Cambodia, 53
Cameroon, 70, 97
Canada, 3, 21, 22, 24, 28, 29 30, 31, 33, 38, 39, 41, 42, 43, 44, 45, 47, 49, 50, 52, 57, 59, 81, 88, 89, 91, 94, 96, 101, 102, 107, 109, 110–13, 115, 122, 123, 211, 216–17, 233, 234; Canadian International Development Agency (CIDA), 94, 96, 216, 242
capital intensity, 27–33, 148, 231, 247
carbon, 4, 10, 81, 103, 137–38, 163
Carter Holt Harvey, 33, 88, 107, 175

CellMark, 88, 215
Cellulose Attisholz, 176
Celulosa Arauco, 171
Celulose Nipo-Brasileira, 59, 62, 149, 159–62
Central African Republic, 97
centralization, 4, 11–12, 14, 16, 18, 27–30, 33–36, 51, 87, 104, 134, 143, 245, 247, 249, 250, 251
certification, 140–43, 248, 250
Champion International, 62, 106–7, 167
charcoal, 147, 155
Chatichai Choonhavan, 237
Chemical Bank, 217
Cheng Loong, 112
Chile, 3, 5, 6, 9, 12, 15, 29, 41, 43, 44, 45, 47, 48, 49, 50, 52, 56, 57, 58, 59, 60, 65, 66, 67, 70, 72, 73, 83, 85, 88, 94, 101, 107, 109, 111, 113, 131, 136, 169–85; 192; Corporation for the Production of Promotion (CORFO), 171, 173, 180
China, 5, 8, 9, 14, 24, 42, 44, 45, 46, 47, 54, 57, 59, 97, 107, 109, 111, 113, 163, 214, 217, 218, 241, 247
Chipdeco, 215
chlorine, 3, 20, 23, 35, 74, 81, 82, 92, 117, 118, 124, 154, 160, 163, 166, 224, 248
Chuetsu Pulp, 33, 112, 159
Chung Hwa, 217
Confederation of Independent States (CIS), 8, 42, 57
Citibank, 155, 175
Citicorp, 176
Citizens Clearinghouse on Toxic Wastes, 253
Clark, David, 32, 36, 118–19
clearcutting, 107, 111, 134, 141, 180, 220, 221, 225
climate, 3, 65, 84, 119, 136–9, 188, 189, 190
Compañía Manufacturera de Papeles y Cartones (CMPC), 175–76
Colombia, 7, 109
commons, industrial plantations as threat to, 13–15, 82, 105, 136, 220, 222, 224, 226–28, 234–38, 243–45, 249
Companhia Vale do Rio Doce (CVRD), 59, 155, 156, 159, 160, 167, 168
concentration: of land, 13, 32, 154, 157, 158, 160, 168, 195, 202; of ownership, 28, 48, 211–13; of power, 84, 85, 161, 168, 172; of production, 32–4, 48, 172, 211–13; of wealth, 84, 85, 172
Congo, 7, 9, 14, 38, 47, 88, 97
conifers, 3, 6, 8, 9, 17, 48, 49, 50, 54, 58, 72, 147, 148, 230. *See also* pine
consultants, 15, 27, 32, 48, 52, 53, 71, 89–91, 93–94, 96–100, 104, 119–20, 125, 129, 139, 141, 143, 149, 174, 211, 215, 216–17, 218, 222, 224, 231, 233–34, 241–45, 251, 252

Consultative Group on International Agricultural Research (CGIAR), 100; Center for International Forestry Research (CIFOR), 100
consumption, 10, 11, 12, 14, 18, 23–27, 29, 30, 31, 33, 35, 36–40, 41, 47–8, 54–9, 85, 92, 116, 117, 118–19, 120, 122, 125, 131, 133–4, 136–37, 147, 149, 170, 176, 200, 213, 228, 229, 230, 232, 238, 243, 244, 247–49; increases in, 17, 37, 39; world distribution of, 37–39, 133–34; driven by packaging and other business uses, 37–38, 132–33; of chlorine-free pulps, 20, 117, 154, 160, 163, 166
contract farming, 114–15, 156, 161, 224, 230, 239–40
Compañía de Petróleos de Chile, 174, 176
Costa Rica, 7, 14, 38, 109, 137, 149
Côte d'Ivoire, 96
Cranbrook, Lord, 143
Crecex, 179
Creusot-Loire Enterprises, 172
crops, agricultural, 5, 13, 67, 68, 83, 135, 141, 170, 174, 177, 183, 195, 202, 204, 239–40, 247; comparison of plantations with, 60, 61, 76, 85, 132, 205, 208–09, 233
Cruzat-Larrain, 176
Commonwealth Scientific and Industrial Research Organization (CSIRO), 104
Cuba, 7
currency fluctuations, 30–31

Daiko Paper, 159
Daio Paper, 110–11, 159, 175
Daishowa Paper, 31, 46, 59, 102, 108–09, 159
dams, hydroelectric, 65, 191, 241
David S. Smith, 112–13
deforestation, 10, 49, 61, 140, 147, 155; industrial tree plantations as cause of, 13, 67, 70, 82, 131–32, 138, 152–53, 158, 160, 161, 162, 167, 170, 178, 180–82, 208, 211, 217, 219–23, 226–27, 235–38, 241
degraded land, 63, 227; inability of industrial tree plantations to rehabilitate, 63–66, 75–80, 82, 134–36, 153, 161, 183, 205, 219, 232, 237
Denmark, 24
Dias Leite, Antonio, 150
dictatorship, 147, 152, 162, 168, 169, 172–75, 179, 185, 237
dioxins, 20, 26, 81, 102, 114, 252–53
disease, 51, 69, 72, 73, 74, 150, 157, 163, 175, 187, 194, 225
dispossession, 57, 84, 152–54, 157, 160, 167, 177–78, 203, 209, 222, 225–27, 235–37, 239, 252

divide and conquer, 84, 114, 119, 121, 125–26, 127–43, 225–26, 240, 251
Djamaloeddin Soeryohadikoesoemo, 226
Dominican Republic, 14, 38, 120
Dow Chemical, 91, 123, 207
drought as exacerbated by industrial plantations, 65, 184, 208, 209, 225
Duchin, Ronald A., 118

ECOTEC consultants, 149
Ecuador, 7, 137
efficiency, 138; industry lack of, 27–28, 102–03, 136
Egypt, 16, 38, 214
Ehrnrooth, Henrik, 120
Eka Nobel, 91
Eka Tjipta Widjaja (Oei Ek Tjhong), 213
Elof Hansson, 215
employment, 82, 84, 189, 195, 198, 224; few jobs on industrial plantations, 15, 84, 86, 139–40, 149, 152, 157, 161, 164–65, 177–79, 184, 202–03, 235; few jobs in pulp and paper mills, 29. *See also* labour
ENCE (Spain), 195
energy, 3, 12, 19, 26, 75, 137–39, 156, 158, 159, 161, 163–64, 201, 210, 215
Enso, 33, 46, 106, 119, 217, 219; ENSO/Indufor, 89
environmentalism, 14, 15, 20, 23, 35–36, 48, 54, 56, 60, 61, 62, 76, 91, 92, 107, 119, 121–26, 130, 131, 134, 137, 140, 141, 142, 143, 151, 160, 163, 165, 166, 192, 229, 236, 237, 243, 248
erosion, 76, 170, 191, 193; caused by industrial plantations, 45, 78, 80, 164, 183, 184, 209, 227, 235
Ethiopia, 7, 8, 97
eucalyptus, 3, 5, 6, 13–14, 15, 18, 26, 47, 49, 50, 51, 56, 58, 60, 61, 62, 63, 65, 66–67, 68, 70, 71, 73, 74, 75, 76, 77, 78, 79, 80, 83, 85, 86, 94, 95, 96, 97, 103, 107, 109, 111, 116, 132, 135, 141, 147–56, 158–63, 166–68, 170, 174, 175, 176, 178, 180, 185, 186–87, 188, 191, 194, 198–99, 201, 220, 225–26, 230–41, 245; *Eucalyptus camaldulensis*, 6, 83, 186, 231, 235; *E. cinerea*, 186; *E. diversicolor*, 186; *E. globulus*, 6, 174, 186–87, 194, 198; *E. grandis*, 6, 194, 199; *E. robusta*, 6, 186; *E. saligna*, 6, 78, 79, 186; *E. tereticornis*, 6, 186
European Overseas Development Corporation, 230, 240
export trade, 12, 18, 21, 23, 31–36, 41–59, 84, 88, 98, 104, 113, 122, 125, 134–5, 140, 143, 148, 149, 155, 160, 163, 170, 174–76, 188, 191, 195, 200, 213–18, 229, 230; main exporters of fibre, pulp and paper, 43–45; main Southern exporters of fibre and pulp, 147–85, 198–228; main importers, 46–47
export credit agencies, 95–96, 219
Export-Import Bank (Japan) 96

Fabrica Nacional de Papel (Uruguay), 60
Fearnside, Philip M., 71, 135, 138–39, 165
Ferguson, James, 245
Fernandez Carro, O., 51, 121, 130
fertilizers, 62, 69, 72, 75, 79, 80, 158, 161, 164, 182, 184, 187, 188, 190, 241
fibre, cellulose, 16; long, 17, 18–20, 22–23, 26, 31, 35, 47, 55, 56, 148, 163, 230; short, 17, 18–20, 23, 35, 47–48, 56, 58, 148, 163, 191, 199, 230; production and trade, 41–47
Fiji, 7, 42, 57, 96, 109
finance, 29, 30, 93–100, 102, 134–35, 141, 148, 150, 159, 165, 171, 173, 189, 191, 196, 200, 204, 215–19, 226, 228, 230, 232, 234, 252
Finland, 3, 21, 22, 24, 25, 28, 33, 38, 41, 42, 43, 44, 45, 46, 47, 53, 89, 90, 91, 94, 96, 101, 106, 107, 109, 110, 119, 120, 122, 192, 211, 215, 216, 217, 219, 224, 233, 234, 240, 242, 245, 252; FINNIDA, 94, 216, 217, 242, 243, 244, 252; Finnish Export Credit, 216; Finnish Fund for Industrial Cooperation, 217; Premixed Concessional Credit Scheme, 53, 96
fire, 72, 73, 177, 178, 180, 182, 194, 201, 204, 206, 207, 209, 222, 223
fisheries, industry threats to, 66, 81, 153, 157–58, 162, 164, 166, 184, 206, 223–24, 226, 235
Fletcher, Hugh, 30
Fletcher Challenge, 30, 46, 108
FLONIBRA, 159, 161
flooding, 183, 184, 190
FLORAM (Brazil), 88
Florestas Rio Doce, 159
Foelkel, Celso, 166
Food and Agriculture Organization (FAO), 13, 14, 21, 25–26, 35, 39–40, 54, 68, 71, 75–76, 80, 99, 100, 133, 162, 171, 174, 189, 194, 198, 214, 235, 242, 243
FORENCO Consultants, 89, 241
Forestal Ace, 175
Forestal Arauco, 174
Forestal Oriental, 191, 192, 193
Forestal Santa Fe, 176
foresters and forestry, 8, 10, 11, 13–14, 30, 32, 34, 53, 60–63, 66–82, 85, 100–01, 103–04, 116, 126–30, 136, 140–43, 149, 164, 180, 187, 189–90, 193, 198, 207, 241–45, 246–48, 251, 253
Forestry Institute (Chile), 180
Forestry Stewardship Council, 140

France, 16, 21, 22, 39, 42, 43, 46, 88, 89, 91, 99, 107, 109
Freund, Alfred, 166
Friends of the Earth, 247
fruit, 153, 157, 178, 236
fuel, 187, 191, 194, 201, 210, 235, 236
fungi, 72–74, 77, 83, 180–82, 194, 236, 241

Gadgil, Madhav, 28
Galbraith, John Kenneth, 36
General Agreement on Tariffs and Trade (GATT), 34, 52
Georgia-Pacific, 106–07, 213
Gerardt, Artur, 152
Germany, 16, 22, 24, 33, 38, 39, 42, 43, 44, 46, 47, 50, 54, 82, 89, 91, 94, 98, 107, 109, 110, 112, 113, 166, 200, 211, 213, 215; GtZ, 94
Ghana, 7, 38, 97
global warming, 4, 124, 137–39, 163–64
globalization, 30–36, 37–38, 41–59, 86, 104–5, 114, 126, 128, 133–34, 143, 200
glyphosate, 153, 207, 241
gmelina, 6, 74, 79, 137, 162, 163, 164
Gonzalez Castillo, Jose, 179
Graham, Alastair, 33
grasslands, 14, 64–65, 68, 71, 82, 186–97, 204, 205, 207–08, 217, 219
Green Revolution, 51, 61, 75–76, 100, 139–40, 183, 188
Grey/MediaCom, 122
growth rates, 6, 9–11, 14, 49, 77, 134, 135, 138, 149, 172, 182, 224
Grupo Vial, 174, 176
Guatemala, 137
Gudung Garam, 217
Guha, Ramachandra, 28

H. A. Simons, 50, 89, 94, 216, 234
Hagler, Robert, 48, 50, 52
Hannover Papier, 200
Hansol Paper, 110–11
Harn Leenanonda, 242
harrassment, 114, 124
Hasjrul Harahap, 219
herbicides, 4, 69, 70, 72, 74, 75, 78, 107, 153, 158, 162, 178, 181, 184, 188, 204, 207, 241
Hiang Seng, 230
Hokuetsu, 59, 112, 159, 215
Honduras, 88, 109
Hong Kong, 38, 46, 111, 113, 213, 214
Honshu, 33, 56, 59, 110, 159
Hutan Tanaman Industri (HTI), 212, 219, 224, 226, 227
Huber, Anton, 184
Hungary, 37, 43
Hunt, Leuchars & Hepburn, 199

ICI, 207, 211
imperalism, 8–11, 98–100
India, 5, 7, 8, 12–3, 15, 16, 18, 28, 29, 34, 38, 39, 40, 46, 49, 67, 72, 75, 89, 90, 94, 97, 116, 134, 135, 211, 214, 217, 230, 237, 247; Forest Research Institute, 67
Indonesia, 7, 8, 9, 12, 15, 24, 29, 32, 34, 38, 40, 41, 42, 43, 44, 45, 46, 49–50, 52, 53, 56, 57, 58, 59, 70, 86, 88, 92, 94, 96, 97, 100, 101, 105, 107, 109, 112–3, 116, 119–20, 122, 132, 135, 141, 142, 162, 163, 211–28, 241, 242, 251, 252; Inhutani III firm, 217, 219
Informes y Proyectos S.A., 193
insects, 73, 83, 181, 194, 207, 208, 221, 236
Institute of Applied Systems Analysis (IASA), 99
integrated mills, 21, 59, 149
Inter-American Development Bank, 97, 167, 174
International Institute for Environment and Development (IIED), 103, 129, 142, 151
International Paper, 32, 33, 52, 88, 106–07, 175
investment, 14, 27–28, 30–34, 57, 61–62, 67, 73, 75, 84, 86, 88–9, 95–97, 100–02, 104, 109, 114, 117, 119, 125, 139, 142, 149–50, 155, 157, 164, 166, 168, 171–75, 188–89, 190–92, 194–97, 211, 214–15, 217, 219, 223, 228–30, 234, 237–38, 240–41, 243, 247, 252
Ireland, 38, 108
Italy, 14, 22, 38, 39, 44, 46, 99, 107, 110, 200
Itochu, 59, 88, 159, 175, 230
International Timber Trade Organization (ITTO), 140, 142

Jaakko Pöyry, 49, 89, 90, 94, 100, 103, 119, 122, 131, 133, 141, 211, 216, 218, 222, 234, 240, 241–45; Interforest, 89
James River, 108–9
Japan, 9, 17, 21, 22, 24, 29, 31, 33, 36, 38, 39, 40, 41, 42, 44, 46, 47, 50, 52, 54–59, 88, 91, 92, 94, 96, 99, 101, 102, 104, 106, 107, 108, 110, 111, 112, 117, 120, 134, 142, 149, 159, 162, 163, 167, 170, 174, 175, 176, 190, 200, 211, 212, 213, 214, 215, 228, 230, 233, 234, 241; Japan International Cooperation Agency (JICA), 63, 94, 190, 191, 233; Overseas Economic Cooperation Fund (OECF), 59, 94, 159, 212, 215, 228, 233
Japan Paper Association, 92
Japan Sumatra Pulp, 215
Japan-Brazil Pulp Resources Development Company (JBP), 159
Jari and Companhia Florestal Monte Dourado, 138, 162–65

Jayanti Group, 223
Jefferson Smurfit, 33, 108–09
Jørgensen, Anders Baltzer, 100
Jujo, 33, 56, 59, 107, 159, 215

Kanzaki Paper, 33, 56, 59, 109, 159
Karnataka Pulpwood Inc. (India), 116
Kayu Lapis Indonesia, 216
Keil, Christian, 90
Kenya, 7, 9, 38, 72, 81, 94, 97
Kertas Kraft Aceh, 213, 220
Kiani Kertas 213, 215, 217, 220
Kimberly-Clark, 33, 108, 122, 230
Kitti Damnoencharnwanit, 95
Klabin, 62, 112, 165, 166
Kone, 91, 211, 215
Kværner Pulping, 91, 215, 217, 218

Labat Anderson, 224
labour, 3, 50, 51, 84, 101–02, 135, 152, 154,
    156, 157, 160, 165, 172, 178, 179, 180,
    195, 202, 203, 204, 211, 212, 216, 218,
    220, 222, 223, 236, 252; costs, 50, 101,
    203, 223; poor working conditions on
    industrial plantations, 152, 157, 161,
    164, 172, 178, 179, 195, 203; low wages
    on plantations, 152, 161, 178, 179, 195,
    203, 223, 241. *See also* employment
Läitaläinen, Rauno, 243–5
land, 3, 4, 5, 8, 11, 13–15, 17, 34, 48–51, 54, 58,
    65, 67, 82–6, 88, 89, 95, 101, 105, 107–13,
    114–17, 119, 131, 133, 134–5, 140–41, 147,
    150–53, 155–61, 165–69, 171–73, 175–78,
    183, 185, 189, 195, 197, 200–04, 207, 209,
    216, 218–20, 222, 224–27, 229, 231–
    33, 236, 238–41, 243, 244, 246, 250, 252;
    common, 13–15, 82, 105, 135, 220, 222,
    224, 226–8, 234–38, 243–45, 249;
    ownership of, 189, 194, 203, 225, 236, 238,
    244; prices of, 49, 155, 188, 223; public,
    49, 56, 82, 101, 113, 116, 173, 175, 198–99,
    218–19, 224, 231–33, 235, 238; redistribu-
    tion of, 173, 202, 204, 224, 244;
    speculation in, 95, 115, 202, 235–36
Landegger, George F., 103
Lao PDR, 38, 49, 53, 90, 97, 235, 240–41
Latorre, Jaime, 193
leaf litter, 64, 70, 73, 77, 180, 183
Leslie, Alf, 35
leucaena, 6, 221
liberalization, 33–34, 52, 169
lignin, 19, 20, 22, 23, 37, 77, 81
Lima, Walter de Paula, 60, 62
Lippmann, Walter, 95
livelihood, industrial plantations' threat to,
    84, 94, 131–32, 151, 153, 157, 165, 177–79,
    184, 201–02, 204, 208, 218, 223, 224, 225,

226, 235, 238, 241, 245, 246, 249. *See also*
    employment
livestock, 66, 67, 83, 139, 178, 184, 186–87,
    189, 202, 204, 227; cattle, 71, 186, 194,
    195, 226
Lobl, Alfredo, 166
logging, 41, 44, 47, 52, 55, 123, 129, 132,
    142–43, 155, 183, 208, 211, 212, 219–
    25, 227, 231, 253
Lorentzen Group, 150
Ludwig, Daniel, 162, 164
Luxembourg, 44, 47

machinery firms 90–1, 215–18
MacMillan Bloedel, 28, 59, 110–11, 122
Madagascar, 7, 49
Malawi, 7, 72
Malaysia, 7, 8, 38, 39, 54–57, 59, 70, 94, 109,
    113, 131, 137, 142, 212, 214
mangroves, 17, 18, 120, 215
Marubeni, 59, 88, 109, 120, 212, 215, 216
Matte-Alessandri, 176
Mead, 108–9
mechanization, 4, 17, 27–30, 35, 36, 45, 64,
    80, 139, 153, 177, 188, 189, 203
Medeiros, Rogerio, 153
Mexico, 7, 9, 14, 24, 38, 41, 47, 57, 104,
    109, 113, 123
migration, 152, 157, 161, 165, 177, 178, 202,
    209, 212, 219, 220, 221, 223, 236, 237, 244
Mitsubishi Heavy Industries, 91, 92, 211,
    230
Mitsubishi Paper, 59, 112, 159, 170, 175
Mitsui, 59, 109
Mohammad ('Bob') Hasan, 213, 215, 217, 221
Molina, Antonio, 173
Mondi, 54, 112, 199, 200, 201
Mongoven Biscoe & Duchin, 118, 121
monoculture, 3–6, 10–11, 13, 35, 61, 63, 68–
    69, 71, 74–77, 90, 94, 105, 107, 132, 136,
    142, 148, 152, 157, 164, 166, 171–72, 181,
    188, 192, 194, 202, 209, 219, 246, 250
Monsanto, 207
Monte Dourado, 62, 149, 162, 163, 165
Morgan Grenfell Asia, 217
Morocco, 52
Mozambique, 7, 94, 109, 201
Mukerji, A. K., 90
multilateral agencies, 12, 52, 82, 89, 93, 96–
    100, 137, 174, 192, 235, 242
murder, 126, 236
Musi Hutan Persada, 226
Myers, Norman, 104, 137

Narong Mahanond, 232
Nat Inthakan, 242
Natal Agricultural Union, 83, 206

National Association of Pulp and Paper
   Producers (Brazil), 62
National Confederation of Forestry Workers
   (Chile), 179
native forests, 9, 10, 15, 16–18, 35, 54, 56, 70,
   107, 111, 113, 131–32, 135, 137, 141–42,
   151, 154, 156, 158, 160, 161, 163, 165, 169,
   170, 172, 174, 178, 181, 186, 198, 207, 208,
   211, 219–23, 227, 232, 238, 241; compared
   to industrial plantations, 3–5, 66, 71, 77,
   182–84, 235–36
Nepal, 38, 94, 97, 242, 252
Netherlands, the, 24, 25, 46, 47, 107, 108,
   111, 220
New Caledonia, 7
New Oji, 33, 59, 108, 113, 230
New Zealand, 5, 6, 9, 15, 33, 38, 41, 43, 45, 46,
   49, 50, 54–56, 57, 59, 65, 68, 74, 88, 89, 107,
   108, 109, 140, 175, 176, 230, 241;
   Development Assistance Division, 140
non-governmental organizations (NGOs), 51,
   103, 104, 119, 121–22, 124–29, 130, 136,
   143, 150, 151, 153–54, 157, 170, 193, 196,
   223–25, 228, 234, 238, 242–44, 249, 252
North American Free Trade Agreement, 34
Nicaragua, 7, 38
Nigeria, 7, 44, 70, 77, 97
Nippon Paper, 33, 59, 106–07, 175, 212, 215
Nissho Iwai, 167
non-wood fibre crops, 13, 16–8, 26, 232, 247–9
Noordwijk Declaration, 137
Norcell, 62
Norway, 42, 43, 44, 93, 110, 165
nutrients, 5, 70, 75–81, 83, 161, 164, 182,
   187, 194, 235

Organization of American States, 190, 193
Oberlander, Ronald Y., 31
Oji Paper, 33, 59, 109, 111, 113, 159
Olsson, Roger, 25, 40
organochlorines, 81, 166
Orientalism, 129
Otero, Luis, 183
O'Toole, Randall, 103
Outukumpu, 91, 211

packaging, 16, 23, 37–38, 90, 97, 107, 111,
   113, 213, 217, 229
Pakistan, 7, 97, 214
Pan African Paper Mills, 81
Pangestu Prayogo, 212
Panjapol, 96, 230, 238
paper, 16–17, 21–27; board, 23, 25, 37, 107,
   111, 113; coated paper, 20–22, 25, 26,
   200; fillers, 21; ink, 23, 25–27, 33, 35;
   machines, 16–18, 21–23, 27–30, 32–36,
   53; newsprint, 17, 19, 22–23, 25, 27, 30,
   37, 41, 59, 107, 109, 111, 113, 133, 230;
   printing and writing papers, 20–22, 36,
   54, 107, 111, 113, 155; production of, 21–
   26; tissue, 18, 20, 22, 23, 37, 109, 230;
   trade in, 41, 44–47; waste paper, 23–27,
   41, 42, 47, 117, 238, 247
paper companies, 27–8, 33–34, 52, 54, 59,
   87–88, 95–6, 106–113, 165, 176
Papua New Guinea, 7, 15, 18, 38, 56, 57,
   58, 70, 97, 109, 111, 141
Paraguay, 94, 101
Parsons & Whittemore, 88, 102, 171–72
pasture, 3, 9, 83, 89, 135, 169, 172, 175,
   186ff., 204, 225–6
Pasuk Pongpaichit, 103
Perez Arrarte, Carlos, 86, 193
Peru, 7, 38, 94, 97, 137
pesticides, 4, 72, 74, 124, 139, 153–54, 158,
   162, 178, 184, 187, 188, 194, 207, 241
pests, 45, 72, 73, 74, 83, 156, 160, 163,
   181, 187, 194, 222; infestations caused
   by industrial plantations, 83
Philippines, 7, 14, 38, 46, 74, 94, 97, 104,
   214, 230, 242, 252
Phinma Group, 242
Phoenix Pulp & Paper, 31, 62, 230, 238,
   239, 240; Project Green, 240
pine, 5, 6, 7, 13, 14, 15, 46, 49, 56, 60, 61,
   65–68, 72–75, 77, 79, 80, 88, 94, 96, 104,
   107, 147, 160, 163, 168, 171–78, 181–84,
   191, 194, 198–200, 205, 207–08, 214, 215,
   221, 225, 253; *Pinus caribaea*, 6; *P. elliotti*, 6,
   194; *P. merkusii*, 6, 220; *P. patula*, 6, 72, 198,
   208; *P. pinaster*, 72, 186; *P. radiata*, 6, 66, 68,
   72, 74, 169, 170–74, 176, 182, 183, 194,
   198, 205, 241; *P. roxburghii*, 72; *P. taeda*,
   194; *P. wallichiana*, 72
Pinochet, Augusto, 172
Poland, 38, 97
pollution, 3, 20, 26–27, 35, 56, 81, 107, 111,
   137, 148, 156, 158, 160, 162, 163, 165–66,
   183, 184, 187, 196, 247, 252; air, 80, 154,
   226; water, 82, 154, 208, 223, 226, 240,
   248; soil, 240
poplar, 17, 186
Portugal, 5, 9, 15, 38, 42–45, 50, 54, 59,
   107, 109, 111–13, 201
Prem Tinsulanonda, 242
public relations and propaganda, 103, 117–
   19, 122–26, 131, 141, 151, 156–58, 160,
   163–64, 168, 189, 193, 228, 233
pulp, 1, 3, 5, 6, 8, 11, 12, 13, 14, 15, 17, 18, 19–
   35, 40; chemi-thermo-mechanical, 20,
   23, 81, 191; chemical, 19, 20, 22, 80, 174,
   176; kraft, 20–3, 29, 31, 34, 80, 81, 102, 159,
   167, 191, 200, 230; market, 21, 28, 41, 42,
   47, 88, 107, 109, 111, 113, 149, 155, 163,

188, 199, 214; mechanical, *19*, 20, 22, 23, 81; prices for, 26, 30, 31, 32; sulphite, 20, 22, 23

pulp and paper business cycle, 11, 25, 26, 30–36; and risks for the South, 85–6

Rainforest Action Network, 247

Raja Garuda Mas, 46, 211, 212, 214, 217; Indorayon, 212, 214, 215, 217, 220, 223–24, 225–26, 228; Riau Andalan, 211–13, 215, 220, 221

Rasmusson, Ulf, 53

rattan, 16, 226, 227

rayon, 13, 23, 166, 212, 215, 220, 228

recycling, 19, 22, 23–27, 34, 40, 56, 92, 111, 117, 118, 248; limitations of, 19, 25–26, 40, 49, 248

reforestation, 5, 65, 67, 160, 186, 217, 224, 231, 233, 236–38, 249

Rengo, 112

repression, 50, 84, 101, 115, 119, 141, 174, 177, 178, 195, 202, 203, 223, 226, 237

resistance, 12, 15, 49, 54, 67, 71, 72, 103, 104, 105, 107, 109, 111, 114–18, 120, 126, 130, 148, 152–53, 160, 166, 169, 187, 189, 191, 193, 203, 222, 224–28, 229, 231, 232, 235–41, 243–45, 252; management of, 114–43

Resources for the Future, 103

Rigesa, 111

Riocell, 62, 149, 165, 166

Ripasa, 62, 167

Rogers, Lewis 194

Rosoman, Grant, 73, 76, 78

roundwood, 25, 188, 189, 191, 196, 200

Rural Workers' Confederation (Chile), 173

Russia, 32, 35, 38, 41, 42, 43, 48, 54, 55, 96, 107, 109, 111

Rwanda, 7

Safra Group, 150

Sahapattanapibul, 230

Sakai, 222

Sandwell, 89, 216

Santa Fe, 174–75

Sanyo Kokusaku, 33, 59, 107, 159, 215. *See also* Nippon Paper

Sappi, 54, 96, 108, 199, 200

Sargent, Caroline, 65, 151

Sarriopapel y Celulosa, 88

Sarwono Kusumaatmadja, 223

Saudi Arabia, 38, 215

Scott Paper, 33, 109, 175, 176, 228, 252

scrubland, 167, 201

Senegal, 7

Settsu, 59, 112–13

Shell, 10, 14, 80, 88, 103, 115, 117, 122, 125, 134, 140, 142, 175, 176, 192, 195, 197, 237

Shin Ho, 112–13, 230

Shiselweni Forestry Company, 96

Siam Cement Group, 230, 240; Siam Cellulose, 240; Siam Pulp and Paper, 238, 240, 241

Siberia, 35, 46, 50, 57, 96, 99, 109, 131, 253

Siemens, 91, 211

Sigit Harjojudanto, 213

silviculture, 10, 62, 74, 76, 79, 99, 194, 222

Simpson Paper, 176

Sinar Mas, 34, 46, 213–18, 220, 223; Indah Kiat, 46, 112, 134, 213, 215, 217, 220, 222, 223; Tjiwi Kimia, 216; Wirya Karya Sakti, 88, 213, 220

Singapore, 38, 39, 59, 90, 113, 214, 222

Siti Hardiyanti Rukmana, 212

Smith, Adam, 103

Smith, Maureen, 18, 247

soil, 3, 4, 15, 35, 62–70, 72, 73, 75–80, 82, 97, 132, 134, 135, 151, 153, 154, 160, 162, 164, 180, 183, 184, 187, 188, 189, 190, 191, 193, 196, 197, 206, 209, 219, 221, 222, 224, 226, 235, 239, 240, 250; acidification, 77, 80, 161, 182; carbon, 79; ions, 78

Solomon Islands, 96

Soon Hua Seng, 95–96, 229, 230, 232, 234, 237, 239, 240, 241; Advance Agro, 229–30, 239

South Africa, 5, 9, 12, 18, 38, 41, 43, 45, 47, 54, 56, 57, 58, 59, 60, 65, 66, 68, 72, 74, 83, 84, 88, 96, 107, 108, 109, 112, 198–210

South Korea, 22, 24, 38, 39, 44, 46, 47, 52, 54, 110, 111, 112, 113, 163, 200, 213, 214, 230

Souza Cruz, 88, 150

Spain, 5, 9, 14, 15, 16, 29, 38, 39, 42, 43, 45, 50, 59, 67, 78, 79, 83, 88, 109, 111, 113, 169, 186, 192, 193, 195

Spears, John, 242

Sri Lanka, 7, 94, 97, 113, 214

Stedile, Joao Pedro, 153

steel industry, 155, 158

Stolovich, Luis, 193, 196

Stone Container, 14, 106–7

Stora, 106–7

Strom, Harry, 28

subsidies, 3, 12, 13, 28, 32, 35, 45, 50, 51, 85, 86, 88, 91, 93–105, 107, 111, 114, 125, 132, 134, 139, 141, 147, 148, 149–51, 155, 158, 159, 160, 165, 167–68, 169–70, 172–77, 188–89, 195–97, 200, 204, 212, 216–19, 223, 229, 231–35, 237, 239–45, 246, 252

Sudan, 7, 97

Suharto, 212, 213

Sukanto Tanoto, 212

Sumitomo, 59, 113, 175

Sunds Defibrator, 53, 91, 211, 215, 216, 229

Suresh Kilam, 214

Suzano Group, 154, 155

Swaziland, 46, 88, 96, 135, 201
Sweden, 21, 22, 24, 33, 38, 42–47, 52, 53, 88,
    89–92, 94, 96, 106–9, 112–13, 127, 138,
    150, 211, 215–16, 234, 240; SIDA, 94
Swedforest, 89, 234
Swedish Forest Industry Water and Air
    Pollution Research Foundation, 137
Swedish Pulp and Paper Research Institute,
    103
Switzerland, 17, 89, 91, 149, 193, 215

Taiga Rescue Network, 25, 40
Taiwan, 22, 24, 38, 39, 47, 54, 55, 57, 112,
    200, 211, 214, 217, 230
Takahito Mikami, 190–91
Tampella, 53, 91, 211, 217
tannin, 77, 208
Tanzania, 7, 38, 94
TetraPak, 36, 97, 123
Thai Development Research Institute, 234
Thai-Japan Reforestation and Wood
    Industry Co., 233
Thailand, 6, 7, 12, 15, 24, 32, 37–38, 40, 42, 45,
    46, 49, 52, 53, 57, 58, 59, 62, 66, 67, 71, 72,
    83, 85, 88, 90, 94, 95, 96, 101, 103, 109, 111,
    113, 115, 116, 117, 119, 126, 131, 135, 141,
    176, 214, 229–245, 249, 251, 252; Board of
    Investment, 101, 233; Forest Industry
    Organization (FIO), 231, 232, 234; Royal
    Forest Department, 231, 232, 242–43;
    Forestry Sector Master Plan, 234, 242–
    45
transmigration, 220, 221, 222, 224, 227
Transpapel (Uruguay), 193
transport, 148, 156, 157, 159, 165, 190,
    195–96, 203, 211, 225, 227, 229, 238
trickery, 114, 119, 124–26, 225, 227, 235
Tropical Forest Action Programme (TFAP),
    174, 242, 243
Tunisia, 18, 97
Tupiniquim, 152–53
Turkey, 44, 107

unions, 50, 92, 107, 123, 126, 157, 161,
    172, 174, 179, 193, 195, 203, 206, 252
Union Camp, 48, 108–09
United Kingdom, 16, 17, 24, 37–38, 44–47,
    54, 88–92, 94–96, 98, 101, 109–13, 122,
    124, 126, 132, 142, 200–01, 204, 228,
    234, 247, 252–53; Commonwealth
    Development Corporation, 95, 96, 200,
    234; Department of Trade and Industry,
    98–99; ODA, 96, 98
United Nations Development Programme
    (UNDP), 99, 174, 189, 242–43
universities, 62, 102, 104, 140, 167, 180
UPM-Kymmene, 33, 106–07, 122, 192, 195

Uruguay, 5, 6, 12, 38, 60, 72–74, 83–84, 94,
    101, 107, 186–197, 251
United States, 5, 8, 9, 14, 15, 17, 18, 24, 25,
    29, 30, 31, 37–39, 41–42, 44, 47–50,
    54–59, 81, 87–94, 96–98, 103, 106–13,
    121–23, 125–26, 149, 160, 162–63, 171,
    176–77, 189, 200, 211–13, 215, 217–18,
    224, 228, 230, 242, 247, 253; Overseas
    Private Investment Corporation, 96;
    USAID, 94
Usutu Pulpwood Company, 88, 96, 201

Valmet, 53, 91, 211, 216
Venezuela, 7, 9, 14, 38, 57
Veracruz Florestal, 70, 167
Viet Nam, 7, 14, 38, 53, 94, 97, 109, 212, 214
violence, 152, 161, 169, 178, 225,
    226, 228, 236, 237
Virta, Jouko, 120, 242, 244
Voith, 91, 211, 215
Votorantim, 62, 165, 166, 167

wages, *see* labour
water, 3–5, 13–14, 16, 19, 20, 26, 52, 67;
    ground, 64, 65, 68, 76, 80, 153, 158, 162,
    184, 194, 205, 208, 235; surface, 66–68,
    72, 81, 152–54, 158, 184, 191, 205–06,
    208, 211, 221, 222, 225, 227, 235; cycle,
    15, 63–65, 69, 70, 132, 183, 187, 205;
    supply, 64, 65, 70, 80, 82, 184, 189, 191,
    226; table, 64, 67, 68, 69, 76, 162
Western, A. W., 27, 247
Westoby, Jack, 100–01, 132–3
Westvaco, 110–11
wetlands, 66, 68, 70, 205, 207, 208, 230
Weyerhaeuser, 46, 56, 108, 124
Wilson, Robert A., 50, 51, 121, 130
Women's Environmental Network, 253
wood: chips, 20, 44, 47, 50, 52, 55, 56, 57, 88,
    94, 107, 120, 132, 170, 175, 176, 200, 214,
    230; prices, 50, 203, 223; hardwoods, 6,
    17, 18, 20, 23, 31, 34, 35, 45, 47, 48, 50, 52,
    56, 57, 58, 68, 70, 102, 103, 107, 131, 142,
    198, 199, 214, 220, 221; softwoods, 17, 18,
    20, 22, 23, 31, 47, 50, 52, 55, 56, 57, 58, 198,
    199, 218, 222
World Bank, 6, 8, 14, 36, 40, 89, 97, 98, 100,
    101, 104, 137, 155, 167, 174, 177, 189, 190,
    191, 192, 193, 216, 221, 222, 241, 242
World Resources Institute, 103, 127, 134, 242

Yuen Foong Yu, 217
Zaire, 7, 94
Zambia, 7, 94
Zeneca Agrochemicals, 204
Zerner, Charles, 222
Zimbabwe, 7, 9, 97